...aperbacks

...Sc FZS

The Animal Kingdom

illustrated by David Pratt

Hamlyn · London
Sun Books · Melbourne

FOREWORD

With the widening public interest in wild animals and their conservation, aided by the media of television and films, we are made much more aware of the variety of animal life in the world. But many people are still under the misapprehension that only warm-blooded, furry beasts and man are *animals*. This would leave the majority of living organisms unclassified!

So, this book sets out to describe examples of the whole range of animals from the smallest protozoan to man. Each animal is grouped with its nearest relatives, and their groups are outlined. The range of habitats, the divergence within each group, the evolutionary trends, the world wide variation, and some basic zoological principles are described. Of even greater importance, the relationship of species to man and his domestic animals and plant crops, either for their benefit or to their detriment, are stressed.

It is hoped that this little guide to the animal kingdom will widen its readers' understanding and stimulate further interest in all animals, and provide the stepping stones to more detailed and advanced books on the subject.   S.M.

Published by The Hamlyn Publishing Group Limited
London · New York · Sydney · Toronto
Hamlyn House, Feltham, Middlesex, England
In association with Sun Books Pty. Ltd. Melbourne.

SBN 600 00081 8

Photoset by BAS Printers Limited, Wallop, Hampshire
Colour separations by Schwitter Limited, Zurich
Printed in England by Sir Joseph Causton & Sons Limited

CONTENTS

566

INTRODUCTION

Of all the millions of animals which inhabit the earth, some are small and apparently simple, while others are large and complex in structure. It might seem that the small, microscopic animals are the simplest living organisms, and in some ways they are, but it must not be forgotten that life has existed on this planet for about two thousand million years, and all the animals which we find today have withstood the test of time. Animals which can master changing conditions are said to be adaptable. Any structure they develop which is useful in the changed conditions is an adaptation. The particular place where they live is their habitat, and the sum total of the conditions in which they live is the environment.

Each different basic type of animal is called a species. The most commonly accepted biological definition of a species is that it is a group of animals which form a viable breeding unit. The offspring produced by the mating of the parents must be able to reproduce more animals like themselves. A common example of a species is the domestic cat; and like every other animal it has been given a scientific name to distinguish it from all other species – *Felis domesticus*. The name *Felis* applies to several species which are grouped together to form a genus. Other species bearing the generic name *Felis* include the European Wild Cat *Felis silvestris,* the Bobcat *Felis rupa* and the Puma *Felis concolor*. Thus a genus is a group of closely related animals denoted by the first name given. This system of calling every animal by a generic and specific name is known as the binomial system and was first established by Linnaeus (Carl con Linné), the eighteenth-century Swedish naturalist. All the names are in Latin and are universal in the scientific world.

Further grouping of the animals continues such that the genera are grouped together as families, whose typical name ending is -idae, for example the cat family is the Felidae.

1 The diverse forms within the animal kingdom : *a. Amoeba,*
b. abyssal fish, *c.* Corn Snake, *d.* Red-breasted Toucan, *e.* Dolphin,
f. Bobcat, *g.* Macaque Monkey skeleton, *h.* skeleton of Man

1

e

a

d

h

b

c

f

g

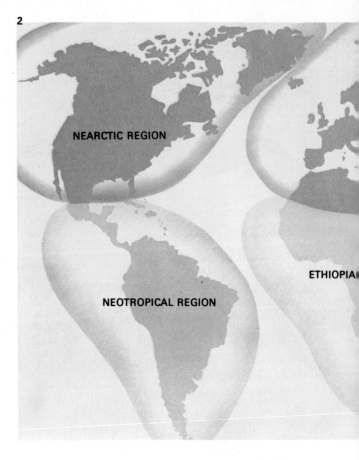

NEARCTIC REGION

NEOTROPICAL REGION

ETHIOPIA

2 The zoogeographical regions

Families are grouped together to give orders, orders to give classes, and classes to give phyla. As will be evident from the following pages, the phylum is a major unit in the zoological filing system. There are some nine major phyla, which embrace the well-known animals such as the Arthropoda and the Mollusca and another thirteen or fourteen minor phyla, depending on which classification scheme is used, including lesser known and fewer species of animals.

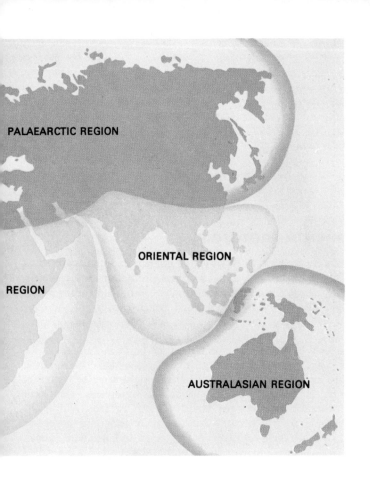

PALAEARCTIC REGION

ORIENTAL REGION

REGION

AUSTRALASIAN REGION

The whole animal kingdom is divided into three sub-kingdoms: the Protozoa, which do not have cellular differentiation; the Parazoa, which simply includes the sponges, isolated by their peculiar grade of organization; and the Metazoa, or multicellular animals, which show an increasing complexity and an evolutionary pattern. This pattern is not clearly defined in some instances, and many transitional stages which were unstable and inadequately adapted must have been lost throughout animal evolution.

3

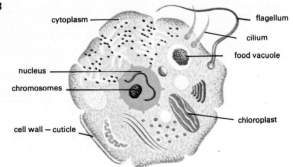

cytoplasm — flagellum

cilium

food vacuole

nucleus —

chromosomes —

chloroplast

cell wall — cuticle

3 Structure of a protozoan

PHYLUM PROTOZOA

The living material of which all animals and plants are made is protoplasm. The cell is a unit of protoplasm and typically consists of a nucleus, which is the central controlling body, and the surrounding cytoplasm which carries out the day-to-day living functions of the cell. The vast majority of animals are multicellular, being composed of many cells which have become specialized to perform functions, for example muscle cells are capable of contraction. The protozoa are unicellular animals, and at first sight seem to represent just one cell, having a nucleus and cytoplasm. On closer inspection, many other structures can be seen within the cytoplasm, which carry out the many functions necessary to enable the protozoan to live as an independent animal. These microscopic structures within the cytoplasm are called organelles [3].

The Protozoa are classed as a subkingdom within the kingdom of animals, because they are so completely different from other animals in that they have no cellular structure. There are four main groups, or classes, into which they are divided based on the type of locomotory structures present and on the methods of reproduction adopted.

Organisms, which move simply by extending portions of their protoplasm into long projections forming 'false-feet' or pseudopodia, are called the **Sarcodina**. Some organisms have long filamentous organelles, flagellae, like whiplashes at one end of the body, and by beating them, drag themselves through the water. These are the **Mastigophora**, alternatively called the **Flagellata**. The **Ciliata** are covered with fine, hair-like organelles termed cilia, and by their co-ordinated beating, they paddle through the water.

All the members of the fourth group are parasitic, and have a special method of reproduction called sporulation. This involves the multiple division of the parent cell to produce numerous progeny. The group is aptly designated the **Sporozoa**.

Most Protozoa are microscopic, varying in colour, from browns and greens to reds and oranges. Most are free moving, but a few form colonies which are visible with the naked eye. The Protozoa are found in very varied habitats: in the sea; in freshwater; in the moisture film on plants and on soil particles; as parasites in the bodies of plants and animals, including other Protozoa.

Types of Protozoa
4 *Amoeba*
5 *Mastigamoeba*
6 *Bodo*
7 *Entodinium*
8 *Arcella*
9 *Codosiga*

The sarcodines

The adult forms of the sarcodines all use pseudopodia to capture their prey and, in the more active forms, they are also used for locomotion. They have a single nucleus which controls the activities of the animal, and the fresh water forms possess contractile vacuoles which remove excess water from the protoplasm and push it out of the animal's body at intervals. Pseudopodia are temporary extensions of cytoplasm and shapes may vary in different species.

The two most numerous groups are the radiolarians [10] and the foraminiferans [12], both of which secrete skeletons. The radiolarians are marine, planktonic, drifting with other pelagic life in the ocean currents, feeding on other protozoans and microscopic plant life. They are distinguished by the presence of silica or strontium sulphate spicules as an internal skeleton, often built into beautiful, intricate patterns.

The foraminiferans are both marine and freshwater creatures, building shells heavily laden with calcium carbonate. The shells are often perforated by miniature pores which permit the extrusion of cytoplasm as pseudopodia. *Globigerina* is one of the most common foraminiferans, it is pelagic, with a spiralled series of diminishing chambers

10 Radiolarians
11 Silica spicules of dead radiolarians collect on the sea floor as ooze. This can be compacted to sedimentary rock and sometimes land masses. The greatest thickness known is 2000 ft.

12

covered with tiny pores. They are important geologically, not only as the major builders of limestone and chalk rock, but as indices in Eocene and Oligocene rocks where different species of *Nummulites* are indicative of particular strata. Some species are also keys to past climates, as their numbers fluctuate according to small temperature changes.

The Amoebida include the commonly known genus *Amoeba* [4] and the parasitic amoebae, *Entamoeba*. Several species of *Entamoeba* exist as harmless inhabitants of other animal's bodies. Six species are found in man. *Entamoeba coli* lives in the colon of man where it feeds on bacteria and cell fragments, it is harmless and in some areas about half the population may be infected. *Entamoeba histolytica* [13], however, is pathogenic, being the cause of amoebic dysentry, and is world-wide in its distribution.

Entamoeba gingivalis feeds on white blood cells and bacteria and lives in the mouth of humans, usually between the gums and teeth. It is not pathogenic but is usually indicative of poor dental care, as it occurs more commonly when there are bad, or carious, teeth, and when gums bleed persistently in the condition known as pyorrhea.

12 Foraminiferans
13 *Entamoeba histolytica* can feed on red blood cells, fragments of the gut or bacteria and generally causes no harm. In some cases it invades the lining cells and causes an ulcerous condition and dysentery

13

frontal cirri

c

lip

preoral cirri

meganucleus

contractile vacuole

sensory cirri

anal cirri

anus

a

14

b

canal

meganucleus

contractile vacuole

a

15

b

The ciliates

This is the largest class of the protozoans, they all possess cilia which are co-ordinated beneath the outer pellicle by a system of fibrils. They are the only class to have two nuclei, a macronucleus which controls the vegetative processes of nutrition, metabolism and regeneration, and a micronucleus which seems to control inheritance and sexual reproduction. Cilia may be distributed evenly over the body surface, as in the Holotrichs, restricted to definite tracts, or 'glued' to give bristle-like structures.

Most ciliates are free swimming and solitary, living in fresh or marine water. Sessile, or sedentary, forms occur, such as *Stentor* [15] which sometimes detach themselves and swim after their prey, others, *Vorticella* for example, have long stalks which they can coil up rapidly like a spring. This offers some protection from danger, but not so much as the hard case secreted by some species into which they can withdraw. Few ciliates develop skeletons, but *Coleps* is a genus which has the ability to produce an armour-plating of small calcareous scales on the pellicle surface.

The marine ciliate, *Mesodinium rubrum,* contains in its ectoplasm red pigment granules. When the animal occurs in swarms, the sea appears to be blood red and these 'red tides' are poisonous to fish. Other genera exhibit colour, due to their own pigment, diffraction of light, or the presence of symbiotic algae.

There are also parasites within this class and the vertebrates have their share of harmful species. Freshwater fish may be infected by a highly pathogenic ciliate, responsible for innumerable deaths especially in fish hatcheries. Mammals are frequent hosts to the genus *Balantidium*. *B. coli* inhabits the colon of pigs and man, where it shares a similar zoological position to the nematode worm, *Ascaris*. The Suctoria are ciliates equipped with tentacles, for example *Didinium*, which are used to trap prey and suck out the edible protoplasm, leaving behind the pellicle or exoskeleton.

14 *Stylonichia: a.* as seen under a microscope, *b.* enlarged side view, *c.* ventral view
15 *a. Stentor b.* diagrammatic drawing

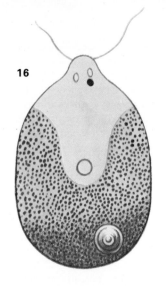

The flagellates

The presence of flagella links together a group of most varied organisms. The range of size in the flagellates alone gives a clue to the variety; some are single, minute, bacteria-sized, others are colonial, jelly embedded forms clearly visible to the naked eye. The group is divided into plant-like flagellates, the Phytomastigina, and animal-like flagellates, the Zoomastigina.

Most flagellates are solitary and free living like the well-studied *Euglena*; others are colonial and still further advance is seen in colonies of *Volvox* [17] with division of labour among the individuals.

The Phystomastigina are usually able to synthesize most of their nutrition requirements from simple substances. The photosynthetic forms may also exhibit amoeboid ingestion and are thus both holozoic, feeding like animals, and holophytic, feeding like plants.

16 *Chlamydomonas*
17 *Volvox*
18 *Ceratium*

14

The Zoomastigina also show a range of types of feeding, taking in whole or partly digested food or actively engulfing food. Parasitic species simply absorb nutrients from their host's blood and tissues. Dinoflagellates possess two flagella which lie in grooves on the body surface. Many species have tough cellulose coats which are ornamented and shaped into the most fantastic patterns, for example *Ceratium* [18] with its tripod form. Phosphorescence of inshore waters is often due to *Noctiluca* [19], and to *Pyrocystis* farther inshore.

The parasitic flagellates include *Trichomonas vaginalis*, frequently the cause of unpleasant irritation in the vagina, *Trichomonas texax*, which inhabits the human mouth, and *T. hominis* the colon. *Giardia lamblia* lives in the human small intestine and causes slight gastro-intestinal symptoms when occurring in large numbers.

Three *Trypanosoma* species [20] inhabit the body of man and two species are the cause of the fatal sleeping sickness commonly found in west to south-east Africa. The Protozoa are transmitted by the tsetse flies, *Glossinia* species.

The plant kingdom is not free of flagellate invaders and *Leptomonas leptovasorum* is known to kill coffee trees in Surinam (Dutch Guiana). All rely on an intermediate insect host which also acts as vector.

19 Adult *Noctiluca*
20 *Trypanosoma*

21 *Anopheles*, the malarial mosquito

The sporozoans

This class of protozoans are entirely parasitic and have life cycles which involve one or more hosts. They reproduce by sexual means, involving gamete formation, and also asexually by multiple fission to give large numbers of sporozoites, or 'spores'. The infective and transmitted stages often have the ability to move and male gametes are often amoeboid to enable them to reach the female gamete and fuse to give the zygote.

The feeding stage is known as the trophozoite. In some species it can divide by multiple fission, to give offspring called schizozoites. These can re-infect the tissues of the same host, or form the reproductive cells. Once the zygote has been formed, it usually has to be transferred to a new host before it can become active again. On entering a new host, the zygote divides again to give numerous trophozoites.

The Gregarines are parasites of annelid worms and insects. They absorb the food and oxygen they need from their host's fluids. *Monocystis* lives in the seminal vesicles of earthworms and *Lankesteria culicis* infects the mosquito which also transmits yellow fever. The sporocysts, or cases containing the spores, are passed out of the host's body and lie in the soil or water, where they may be eaten by un-suspecting earthworms or mosquito larvae.

The Cnidosporidia are the cause of several diseases which affect commercially useful animals. Most attack fish, which develop tumor-like growths as a result of infection, and therefore cannot be used for human consumption. Honey bees are host to the species *Nosema apis*. *Coccidia* are relatively harmless parasites of many invertebrates.

The Malarial parasites are a large group, four species infecting man, and about fifteen known species infecting birds. All require two hosts to complete their life cycle [22], the vertebrate hosts as above, and an invertebrate host, which must be blood sucking [21].

Human malaria is one of the top priority diseases for eradication in the World Health Organization's programme. To date it is estimated that 362 million people live at risk of infection.

22 Life cycle of malaria: Sporozoites enter the red blood cells 1-2. Each forms spores 3-10 which break out at regular intervals 11. Some develop into sexual forms which the mosquito takes in when it bites an infected person 12-13. In the mosquito's stomach these form eggs 14-19 and sperm 20. Each fertilized egg 21 develops into a capsule where new sporozoites form 22-26. These migrate to the salivary glands 27 and are injected as the mosquito bites.
23 Plasmodium of malaria

24

flagella

nucleus

food particles

25

a

b

food vacuoles

meganucleus

Protozoa – summary

The Protozoa, although dealt with first in this book, and usually placed at the beginning of any animal classification scheme, are not the simple, single units of protoplasm that they first appear to be. There is intricate specialization of the cytoplasm into the different organelles whose structures are still not all known, even with the modern aid of the electron microscope. Not only problems of structure, but of sheer numbers and species have yet to be resolved. Not all the Protozoa which exist have been seen, let alone identified.

The emphasis of the last, completely parasitic, class of protozoans may have given an unbalanced picture of the phylum as a whole. It is true that the four classes each contain enemies of man, some serious, like sleeping sickness and malaria which account for millions of lost man-hours of labour each year in the more tropical countries, but at the other end of the scale, the dinoflagellates and the multitude of

24 *Trichonympha*
25 a. *Vorticella*, b. cross-section through *Vorticella*

ocean planktonic forms are, along with the algae, the beginnings of a large proportion of the world's supply of edible protein. These are also a major source of some vitamins which will eventually be available to man.

The fossil forms provide man with valuable building stone as well as adding to his knowledge of geological history. The soil and freshwater species play an incalculable role in breaking down plant and animal debris, so helping the return of carbon dioxide and nitrogen to the air and soil in the natural cycle. Some of the more unpleasant parasites play an important role in a broad biological sense, by helping to control populations, and so maintaining a balance between species.

The Protozoa are exceptionally useful in the field of research for work on genetics, nutritional requirements, metabolic processes of all kinds, growth and reproduction. The non-pathogenic species can be bred in the laboratory under well controlled conditions, and because of the fast reproductive rate, many generations can be studied in a matter of weeks.

The abundance and range of habitats is unsurpassed by any other phylum. The marine Protozoa are estimated to occur between five to six million per litre of sea water when most abundant. This varies according to the conditions and seasonal fluctuations of the populations. Freshwater plankton can occur in similar quantities.

Soil dwelling species, and the encysted stages of harmful species deposited in the soil, are both likely to be washed into drinking supply waters. In any country which has no proper sanitation facilities, or the large scale expensive sewage disposal works, re-infection risks are very high. Chlorination kills many forms, but the cysts of *Entamoeba histolytica* are known to withstand five times the amount of chlorine normally added to household supplies.

The main battle to be fought against the Protozoa is the battle of hygiene, proper sanitation, careful washing, cooking and covering of food and the improvement of public water supplies. The killing of flies as one of man's major opponents, and above all, the education of people would see an end to so much disease.

PHYLUM PORIFERA

The sponges are regarded as the most primitive of the multicellular animals. They show no organization into true tissues or organs, and the cells are capable of independent existence with remarkable regenerative properties. Adult sponges are sessile, attached to rocks or plants, and only the simple ciliated larval stage is capable of active movement.

The body plan is simplest in the first stage after the larva settles down to grow into the typical adult. It is a vase-like form, with a stalk for attachment [31a]. Minute pores over the outer surface allow water to flow into the main central cavity and a larger pore at the apex, the osculum, allows the water to pass out [31b-c]. There are no nerve or sense cells in sponges, but cells have taken on specific functions, though not yet organized into proper tissues.

26 Fan Sponge (Africa)
27 Glass-rope Sponge
28 Venus's Flower-basket
29 Crumb or Bread (Britain)
30 Commercial Sponge (Mediterranean)
31 a. simple sponge, b. longitudinal section, c. enlarged section of body wall

26

28

27

29

30

20

waste matter exhaled

31

food and oxygen taken in through body wall

pore

collar cells wave flagella to create water current through pore

pore

a   b   c

The skeletal elements of sponges are used to classify the group still further. The Calcarea have spicules made of calcium carbonate, the Hexactinellida have purely siliceous six-rayed spicules, and the Demospongiae have either siliceous spicules of a different pattern to the above, or spicules with cementing fibres of a horny substance called spongin, or just spongin alone, or no skeleton at all.

All sponges rely on the incurrent water to bring oxygen and suspended food particles. In still water, sponges tend to become elongated, so that the osculum will shoot the stale water a long way from the sponge. In moving water the body form tends to be a low, short structure.

The calcareous sponges are world wide in distribution being mainly restricted to shallow coastal waters. *Leucosolenia, Sycon* and *Grantia* are frequently found on the seashore.

The hexactinellids, or glass sponges, chiefly live at depths of 1500 to 3000 feet, greater numbers and genera being found in tropical waters. The spicules give a beautiful lattice work of opaline silica. Venus's Flower-basket, *Euplectella* [28], is a tall, vase shape. *Hyalonema*, the Glass-rope Sponge [27], appears to be a twisted mass of fine silken threads.

The Demospongiae is the most varied of the three classes. One family, the Spongillidae, are freshwater sponges found in most lakes, ponds and streams throughout the world.

The common bath sponge, found mainly in the Gulf of Mexico, the Caribbean and the Mediterranean [30], has a skeleton of spongin only. *Cliona* is a yellow, boring sponge, which damages corals and mollusc shells. The largest of all sponges is the Loggerhead, which grows in tropical waters.

PHYLUM COELENTERATA

These multicellular animals have a body wall of two distinct layers. The body plan is radially symmetrical, with a single aperture acting as mouth and arms to a simple sac-like body.

The outer layer of cells, the ectoderm, is connected to the inner layer, the endoderm, by a jelly-like layer, the mesoglea. This may be thin, as in the sedentary polyp forms, or it may be expanded to form the massive floats of the jelly-fishes. Embedded in the mesoglea are the nerve cells, which form a nerve net thoughout the body. Coelenterates commonly have the ability to bud, forming offshoot individuals that can either separate and become independent, or remain attached and form a colony. Sexual reproduction, involving the formation of gametes, always produces a simple ciliated larva, a planula, similar to the larval stages of sponges.

The subphylum Ctenophora is a small group consisting of the comb jellies only; they have no distinct polyp and medusa stages, but are solitary, free, pelagic forms. They have no nematocysts, and true muscle cells lie in the mesoglea forming a branching network. *Pleurobranchia* is one of the simplest forms, commonly called the Sea Gooseberry [33 to 37]. It is transparent, oval in shape, with eight rows of combs, each comb is made up of fused cilia which beat in unison. Two tentacles, normally trailed behind the

32 Longitudinal section through the body wall of a hydrozoan.
a. stinging cell undischarged, *b.* sting discharged

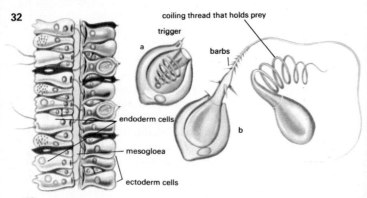

32

coiling thread that holds prey

trigger

a

barbs

endoderm cells

mesogloea

b

ectoderm cells

animal, are covered with special adhesive cells called colloblasts, used for catching prey.

Mertensia is similar to *Pleurobrachia*, but is larger and slightly flattened; both are common in the Atlantic coastal waters. *Beroë* is a larger genus, which is known to act as host to non-photosynthetic flagellates, and exhibit bioluminescence from a chemical source. *Cestus veneris*, Venus' Girdle [38], is about 2 inches wide and nearly a yard long and is found in the surface waters of the Mediterranean. *Gastrodes* is the only known parasitic ctenophore.

The subphylum Cnidaria is divided into three classes—Hydrozoa, Scyphozoa and Actinozoa, based mainly on the relationships of the polypoid and medusoid forms. The ectoderm always bears nematocysts which are of three main types: the glutinants which have sticky hairs, the volvents which coil tightly and the penetrants which are barbed and equipped with hooks for harpooning planktonic organisms.

33-37 Five types of sea gooseberry
38 Venus' Girdle

bud

mouth

young hydra

testis

sperm

ovary

embryo

hydroid

39

a

b

The hydrozoans

The hydrozoans typically have a sedentary, colonial hydroid stage which reproduces sexually to give free-floating buds, the jellyfish-like medusae; these in turn bear the organs of sexual reproduction, the gonads, which liberate the gametes. The zygote, formed by fertilization of the female gamete by the male gamete, develops into a free swimming, ciliated, planula larva which in time settles down, develops tentacles, starts feeding and by budding grows into another hydroid colony [39]. This full life history is detailed in most text books with the genus *Obelia* [42] as a typical example.

The presence of the two forms, the polyp and the medusa, of one animal is known as dimorphism. They are both based on the plan of a mouth surrounded by tentacles, a single enteron or body cavity and two cell layers in the body wall.

In some genera, the polyp becomes the dominant phase of the life history and the medusoid stage is suppressed, for example the hydras, where the polyp produces gonads directly; alternatively the medusoid stage may become dominant as in the Trachylina. The maximum emphasis on the free floating life is found in the siphonophores, *Physalia*, the Portuguese Man-O-War [40], and *Velella*, the By-the-wind Sailor [41], both of which are floating colonies, showing polymorphism of individuals. *Velella* [41] is delicately tinged with violet and blue and *Physalia* [40] is similarly beautiful with iridescent greens and blues on the float, and richer blue tentacles. For all its beauty, however, it produces poison in its nematocysts which will kill small animals and mackerel-size fish and injure humans.

One order of hydrozoans, the Hydrocorallina, are found in tropical and semi-tropical waters as important constituent fauna of coral reefs. *Milliporina* and *Stylasterina* are two genera of hydrozoan corals, they can form low, encrusting growths or tall upright branches depending on the water conditions.

The most common hydroids are small branching colonies found attached to seaweeds and rocks low down on the shore or just offshore. *Sertularia* colonies reach 2 or 3 inches in height and are often cast up on the shore after a rough tide. *Cordylophora* is mainly a brackish water species, and *Tubularia* and the hydras are the few freshwater representatives of the class.

39 Structure and life cycle of *Hydra a.* cross-section through body wall, *b.* life cycle
40 *Physalia*, Portuguese Man-O-War
41 *Velella*, By-the-wind Sailor
42 *Obelia a.* medusa of *Obelia*

The scyphozoans

The term jellyfish is most generally used in reference to this class. They range from the cold Arctic and Antarctic seas to tropical waters, and from the large surface floating forms to the smaller, brilliantly coloured species of ocean depths.

The jellyfish is the dominant medusoid stage, and the hydroid stage is reduced in most to a sessile larva. Gonads are born on the medusa, and after fertilization the zygote develops into a small, two layered, ciliated, typical planula larva. This settles on a suitable rock surface and forms the scyphistoma. The function of the scyphistoma is to multiply asexually. The first stage is polypoid and looks rather like a hydra, then a series of saucer-shaped buds form behind its crown of tentacles, each developing rounded tentacles. When complete, the buds separate and float off singly as an ephyra larva. Each ephyra is a minute larva possessing a mouth leading to the stomach, and a dome of jelly, or bell. It gradually becomes like the adult, so that by the first summer time it is a sexually mature medusa [43].

The medusae have dome-shaped bells and the gonads can frequently be seen in bright colours. The margin of the bell is fringed with tentacles and the mouth is born in the centre of four 'arms'. The outer edge of bell tissue contains

43 Life cycle of *Aurelia*

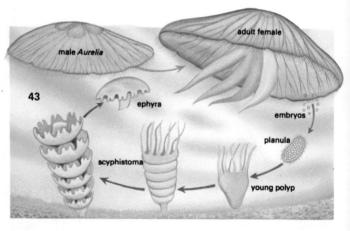

a band of circular muscle which can contract rapidly and shoot the jellyfish up through the water. Some species have simple eyes, or ocelli, as photoreceptors.

Aurelia aurita [43], reaching up to 18 inches across, is the most widely distributed North Atlantic species, its four horse-shoe shaped gonads can be seen pale violet through its colourless float. *Dactylometra quinque-cirrha* has warlike clusters of stinging cells and golden tentacles that partially veil the pink mouth lobes. This species rarely exceeds 8 inches in diameter and is also found in the North Atlantic. A dangerous Pacific species is *D. pacifica* [44] whose lovely appearance belies the danger of its stinging threads. *Rhizostoma* [46] feeds through numerous small openings in the mouth lobes, which hang from the lower surface. There are no tentacles. *Pelagia* has eight long tentacles and is remarkably phosphorescent.

44 *Dactylometra pacifica* (Japan)
45 *Dactylometra*, Sea Nettle (Britain)
46 *Rhizostoma octopus* (Britain)
47 *Cassiopeia: a.* swimming in normal position, *b.* in feeding position with the many-mouthed lobes gathering food particles

48

49

gullet
mesentery
septa
basal disc

The actinozoans

The basic sac-like body of the coelenterates is complicated in the actinozoans by the development of mesentries, sheets of tissue, which partition the enteron into separate cavities. There is no medusoid stage. The sea anemones and stony corals form one order of this class, the Zoantharia. They are distinguished by their mesentries and tentacles occurring in multiples of six and by having simple tentacles [49].

48 Cross-section through budding coral branch
49 Cross-section through anemone
50 Various coral structures
51 Coral formations: *a.* fringing reef, *b.* barrier reef, *c.* atoll
52 Various anemone forms

50

51   a   b   c

Sea anemones are commonly found in coastal waters. They have no skeleton, but a soft column of tissue surmounted by a crown of tentacles. The base of the column forms an attaching disc. The tentacles vary in length, number, fineness, ability to retract and proximity to the mouth in the centre of the oral disc. The largest anemones have been found on the Great Barrier Reef of Australia, *Stoichactis* reaching one yard in diameter across the top, although the majority are only 2 to 3 inches. Several anemones are found as commensals, living on crab claws or shells.

Corals are either solitary or colonial polyps which secrete a calcareous skeleton. Each mesentry is draped over a partition of calcium carbonate, to which it adds throughout life. The partitions, or septa, are fused at the edges to form a coral cup, and in the colonial forms, they are all cemented together to form the large familiar pieces of coral. In the living state, all the surface is covered with soft tissue and polyps; and the colonies usually grow by new polyps budding off the edges. Coral species live in all temperatures of water but the richest development is in the warm oceans up to depths of 3000 feet.

The second order, the Alcyonasia, includes the corals that have mesentries in multiples of eight and tentacles which are divided pinnately so they appear to have a row of small branches on each side of the main axis – like a feather [48]. They include soft bodied colonies of hydroids, such as *Alcyonium digitatum*, Dead Man's Fingers, the Gorgonid corals, of tree-like density and a wealth of colour and the rich red pipes of the Organ-Pipe coral, *Tubipora*.

PHYLUM PLATYHELMINTHES

The flatworms are the first group of animals dealt with to have three true layers of cells, first formed in the embryo and modified in the adult. The outer layer, the ectoderm, forms the protective layer next to the environment, and is sometimes used to help locomotion if ciliated. The inner layer, the endoderm, forms the lining of the gut, and is often able to produce enzymes which digest the food material. The flat worms also have a proper layer of cells, the mesoderm, separating the endoderm and ectoderm, which has the ability to develop into different tissue types.

The new versatility of the mesoderm shows itself in the sudden appearance of clearly defined 'systems' [54]. There is an excretory system. The nerve cells are amalgamated and co-ordinated in a limited way to form a nervous system. The scattered testes are joined up by canals; the ovaries are closely connected by the uterus, the total is the reproductive system. Another advance of the flatworms beyond the coelenterates is that radial symmetry, regarded as primitive, has been lost, and the creatures are bilaterally symmetrical.

As the name suggests the platyhelminthes are nearly all flattened dorso-ventrally, so that there is very little distance

53 Flatworm, *Planaria* species

54 Cross-section through sexually mature *Planaria* showing internal organs. Gland cells open to the surface.

testis

excretory pore

oviduct

yolk gland

dorsal surface

nerve chord

cilia

54

between the back and front surfaces. Typically they have a mouth near the anterior end, opening on to the ventral surface, which leads to a gut made up of branched pouches, or caeca; they have no anus. The extremely modified parasitic forms have lost all trace of mouth and gut.

There are three classes of Platyhelminthes; the Turbellaria, the Trematoda and the Cestoda. The turbellarians [56] are the free living flatworms on the whole, though a few species show a gradual adaptation to a parasitic mode of life. The trematodes are all parasitic but they have not become very modified with respect to body structure. They retain the gut, and yet they are highly successful parasites with complex life histories. The cestodes are extremely modified from the basic platyhelminth plan; they are all internal parasites, they have lost all traces of gut and many have adopted a type of budding, or strobilization, as a method of increasing their breeding potential. So the three classes show the gradual progression and modification from an entirely free living flatworm to an endoparasitic flatworm that is geared for mass production.

Turbellaria

Most members of this class are free living, carnivorous, nonparasitic flatworms, which all retain their gut. They have an ectodermal covering which bears cilia and these are used to bring about locomotion, the typical gliding progression with no obvious active movement of the body.

The body form is always entire, and not divided into proglottides as are the majority of the cestodes, and varies from ovate leaf shapes to long strap-like forms [55]. Many have extensions, rather like tentacles, from the head end and simple eyes can be present, either just in pairs or in rows along the anterior margin. They are further subdivided into five orders on the type of gut branching.

The primitive Acoela have a gut which is packed with interconnecting cells. They are marine types which inhabit the tidal range of the shore, for example, the sand dwelling *Convoluta roscoffensis,* or pelagic types such as *C. henseni.*

The Rhabdocoela have a straight gut. *Microstomum* is a pond dwelling genus and *Fecampia* is one of the class's few parasites, living inside the lobster.

The most common genera belong to the Tricladida, which have three main branches to their gut. The marine form, *Bdelloura* is an ectoparasite, often found attached to the gills of certain crabs. Much more easily found are the freshwater

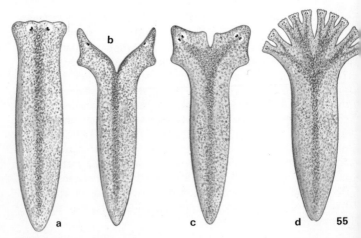

a   b   c   d   55

pond forms, *Planaria* species [53], which have two eyes, and *Polycelis* species which have numerous eyes dotted along the head margin. The terrestrial triclads are the giants of the class, reaching up to 60 centimetres long and being brightly striped on the dorsal surface. *Bipalium kewense* [57] is a tropical form found almost universally in greenhouses. It reaches a foot in length and has an axe-shaped head scattered with eyes.

The Polycladida are entirely marine, a few being commensal with hermit crabs, or various echinoderms. *Stylochus frontalis* is a notorious carnivore which is a pest of oyster beds, eating the soft body of the oyster piece by piece. The last order, the Temnocephalea, are parasitic on freshwater crustaceans. They are equipped with suckers and have tentacles from their anterior end.

55 Capacity for regeneration in a flatworm. If the head *a.* is cut down the middle *b.* each half will regenerate the missing parts *c.* If the head is cut repeatedly and the edges not allowed to grow together, numerous complete heads will result *d.*
56 A turbellarian
57 *Bipalium kewense*, a fresh-water triclad

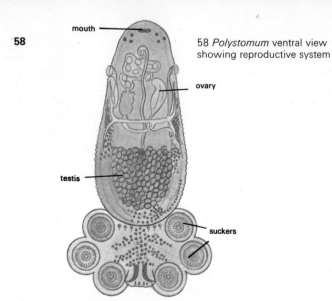

58

mouth

ovary

testis

suckers

58 *Polystomum* ventral view showing reproductive system

59 The Chinese Liver Fluke, *Clonorchis sinensis*, is parasitic in the bile passages of Man's liver. Eggs pass out in human faeces, are eaten by snails and cercariae swim from snail to fish to encyst

59

cercaria encysted in fish develops into adult

cercaria

man eats raw fish

development within the snail

egg

Trematoda

This class of flatworms, commonly known as flukes, are entirely parasitic. They have a protective cuticle covering, a sucker around the mouth, and a variety of posterior or ventral suckers depending on their genus.

The Heterocotylea are dependent on one host only. They attach to the outer surface of their host where it is possible to have easy access to blood and skin tissues. *Polystomium* [58] is the only endoparasitic member of the order, living in the bladder of frogs.

The second order, the Malacocotylea have more than one host; the adult stage is usually parasitic in a vertebrate host, and the larval stages are parasitic mainly on molluscan hosts. *Distomium macrostomum* is parasitic in the gut of thrushes. The eggs pass out with the faeces and are eaten by snails; the multiplicative stages take place and the cercariae, a third larval form, encysts in the snail's brightly coloured tentacles. The thrushes eat the snails and the process begins again.

The commonly cited Liver Fluke of sheep and cattle, *Fasciola hepatica*, frees itself from its molluscan host and encysts on vegetation ready to be consumed by its herbivore host. *Parorchis acanthus* occurs in the rectum of herring gulls that have infected themselves by eating parasitized shellfish, such as mussels or cockles.

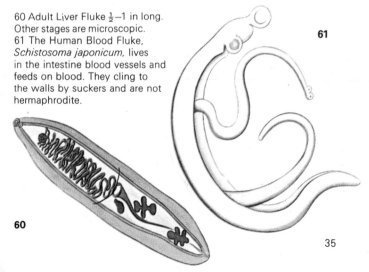

60 Adult Liver Fluke $\frac{1}{2}$–1 in long. Other stages are microscopic.
61 The Human Blood Fluke, *Schistosoma japonicum*, lives in the intestine blood vessels and feeds on blood. They cling to the walls by suckers and are not hermaphrodite.

61

60

62 Man is the intermediate host for *Echinococcus a*. adult lives on the dog, $\frac{1}{8}-\frac{1}{4}$ in long, *b*. cyst from liver of Man

Cestoda

The tapeworms are entirely endoparasitic, with all their members inhabiting the inner cavities of their hosts' bodies in the adult form. They rely on absorption of nutrient fluids from their host through their body wall and have developed an even thicker protective cuticle than the trematodes. An encysted embryo relies on the closely linked activities of the hosts in order to bring about infection.

The first order, the Cestodaria, are small worms which rather resemble the flukes in outward appearance. Most species are parasites which inhabit the gut of fishes. The eggs are released from the adult's body and pass out of the host in its faeces. Should the eggs be eaten, usually by small crustaceans or molluscs, the embryo is released and embeds itself in the tissues of this intermediate host, and is seen to develop further only when the fish eats infected individuals.

The Eucestoda have developed the means of asexual multiplication known as strobilization; the body consists of a clearly defined head, or scolex, behind which is a region which forms buds – strung together to form the tape or

63 Typical section of *Taenia saginata*, Beef Tapeworm
64 Adult Pork Tapeworm, *Taenia solium a.* end of head

strobila. Each bud, or proglottis, develops to a complete set of reproductive organs. By this means, one scolex can produce many reproductive buds and each bud can produce hundreds of eggs, thus greatly increasing the chances of survival of the individual parasite.

Anthobothrium has four stalked suckers on the scolex and parasitizes fish, amphibia and reptiles. Sticklebacks and fish-eating birds are hosts to *Schistocephalus*; pike and man to *Diphyllobothrium*. Adult tapeworms, *Taenia* species [63 and 64], can be found in man and his domestic animals.

PHYLUM NEMATODA

The nematodes, or roundworms, are a group of animals which rival any other group in the animal kingdom in the range of habitat and numbers of individuals. They have conquered the complete range of terrestrial environments and attack almost every group of plants and every class of animal. In general, the nematodes are round in cross-section, having a cylindrical, elongated form, which tapers to a point at both ends. Variations exist in the proportion of diameter to length, in the relative bluntness of the anterior end, in the armature of the head and in the cuticular structures over the body surface. Sense organs are few, but most species bear a pair of amphids on the head, and several free living aquatic species have a pair of simple eyes. The sexes are separate.

Plant nematodes can withstand desiccation, extremes of temperature and an enormous variety of chemical agents which man might apply to crop plants. *Tylenchus tritici*

65

65 Cross-section through a female roundworm, *Rhabditis*

mouth

ovary

anus

66

a

b

c

66 Head structures of nematodes:
a. *Acrobeles*, b. *Wilsonema*,
c. *Seuratia*

infects wheat, the adult developing in the grain head and causing a brown gall to be produced instead of a normal ear. *T. devastrix* attacks a variety of plants: clover, bulbs, onions and other useful plants. Insects act as hosts to many species of nematodes, *Spherularia* parasitizing the Bumble Bee, and *Allantonema* the Bark Beetle.

Man's serious nematode enemies include the trichina worms (*Trichinella* species) that are introduced into the body by eating undercooked meat. The larvae of *Ancylostoma* [67a] and other hookworms bore through man's skin, the adult form eventually causing anaemia. *Dracunculis medinensis* normally lives in copepods and infect man via unfiltered drinking water. The filaria worms (*Wuchereria* species) block the lymphatic ducts of the body causing elephantiasis.

All these nematodes cause various pathological results, all most unpleasant and certainly extremely infectious. Man could well do without his nematode parasites, and without the plant nematodes, which cause such serious economic loss.

67 *a*. Larvae of *Ancylostoma* entering through the skin of a mammal ; *b. Ancylostoma* attached to the small intestine

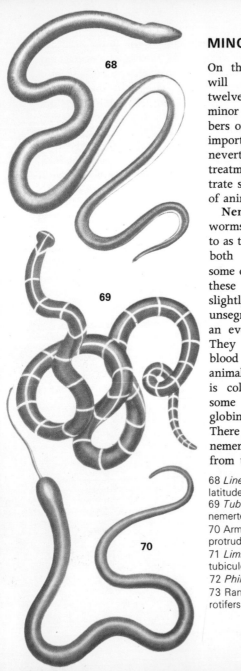

MINOR PHYLA

On these next eight pages will be described some twelve phyla, all regarded as minor with respect to numbers of living species and in importance to man, but nevertheless deserving brief treatment in that they illustrate still further the variety of animal life.

Nemertea, the proboscis worms, may also be referred to as the 'ribbon worms' and both common names give some clue to the structure of these animals. They are slightly flattened, long, thin, unsegmented worms, with an eversible proboscis [70]. They have the first simple blood system found in the animal kingdom. The blood is colourless in most, but some species have haemoglobin in the blood cells. There are some 600 species of nemerteans ranging in size from under an inch to over

68 *Lineus ruber*, a northern latitude nemertean
69 *Tubulanus capistratus*, nemertean from the Pacific coast
70 Armed nemertean with protruded proboscis
71 *Limnias ceratophylli*, a tubiculous rotifer
72 *Philodina roseola*, ventral view
73 Range in body form of the rotifers

60 feet as in the giant ribbon, *Lineus* [68]. Many species are marine and there is one genus of terrestrial species, *Geonemertes*.

Nematomorpha, the hairworms, a small group of some 60 species of extremely fine and elongated worms; the thread-like body is frequently over a foot in length, but rarely is more than $\frac{1}{20}$ inch in diameter, the colour is usually dark brown or black but there are some pale species. The early philosophers thought they arose by spontaneous generation from the hairs of horses' tails.

Rotifera, the wheel organ animals, have a world-wide similarity and are abundant, minute creatures found in ponds and streams. They have a head which bears a crown of cilia, the wheel organ, which is used for locomotion and to create the feeding currents. The trunk carries the main body organs and tapers to a posterior foot. Most species have a protective elastic cuticle which can be ornamented, and others build themselves a tube into which they can withdraw. There are about 1,500 species of rotifers in all, varying in body form and wheel organ structure [73].

71

72

73

74

75

At first glance, the **Gastrotricha** can be confused with rotifers, and the absence of a wheel-organ is the simplest way to distinguish them under a microscope. They are minute, unsegmented, rather worm-like, with a cuticle which is highly specialized; it often forms a flat covering of scales, or an overlapping set of scales rather like a tiled roof. There are some 200 species; the macrodasyoid types [75] are all marine, planktonic and bottom dwellers, and the chaetonotoid types [74] include a few marine species, but mainly consist of freshwater forms which crawl over vegetation and debris at the bottom of ditches and ponds.

Acanthocephala are the spiny-headed worms, a phylum that has only parasitic members. They are adapted to adult life in intestines of vertebrates, and larval life in arthropod intermediate hosts. They have a thin protective cuticle,

76

74 Chaetonotoid gastrotrich, ventral view
75 *Macrodasys*, a macrodasyoid gastrotrich
76 *Priapulus*
77 *Neoechinorhynchus*, an acanthocephalan

through which the digested nutrient fluids intended for the host can be absorbed, and no trace of an alimentary canal. There are about 600 species of acanthocephalians [77]. Man may become accidentally infected by the rat's parasite which reaches up to 10 inches in length if the intermediate insect host, the cockroach, is eaten by mistake.

The mud deposits of shores yield yet another lesser known phylum, the **Kinorhynchia**, consisting of minute creatures, under one millimetre in size. Superficially they bear resemblance to the rotifers and gastrotrichs but they are basically different from either of those two groups in that they have no cilia at all on the body surface. The cuticular covering of the kinorhynchids is ringed giving the appearance of fourteen regular segments. Little is known about them but, so far, there are thought to be about 100 species, these being cosmopolitan, though restricted to mud and silt shore deposits.

The **Priapulida** are represented by five known species of soft-bodied cylindrical worms. They are entirely marine and are restricted to the colder, muddy shores of both hemispheres. They range in body size from 3 to 4 inches, with an anterior proboscis reaching one third of the body length, a central trunk and a series of small pouches forming the caudal appendages. *Priapulus* [76] and *Halicryptus* are the two most commonly found genera.

The Arrow worms **Chaetognatha** are a common feature of plankton samples: they have transparent bodies so that they are difficult to spot in the teeming plankton jar, but their eyes may give them away. They are usually

77

about $\frac{3}{4}$ to $1\frac{1}{4}$ inches long. In the total of 50 species, the body plan follows the same pattern. The head is a short region, with a pair of small eyes, and several pairs of grasping spines on either side of the mouth. The trunk occupies the main length of the animal, with side fins, and the tail bears a broad tail fin. They are all carnivorous. *Sagitta setosa* is typical of coastal waters, *Sagitta elegans* [78], of oceanic waters.

The **Entoprocta** number about 60 species; a few are solitary (*Loxoma*) but the majority are colonial. There is a single freshwater genus, *Urnatella* [84], and the rest are

78 *Sagitta elegans*, an arrow worm, ventral view
79 *Spadella*, dorsal view of a chaetognath that uses adhesive projections to attach itself to objects on the sea floor
80 *Phoronis architecta*
81 Part of *Phoronis hippocrepia* colony
82 Section through an endoproctan
83 Part of a colony of *Pedicellina*, a marine endoproct
84 *Urnatella gracilis*

marine. The colonies are encrusting in form and are increased in size by budding from the growing point of the attaching stalk or stolon. The individuals arise from the stolon by their own stalks and have a polyp-like appearance, with a crown of tentacles at the apex [82].

The solitary *Loxosoma* is found attached to sponges, crabs and sea squirts. When new buds form, they separate off and grow into individuals. *Pedicellina* [83] is a seashore genus, *Urnatella* [84], a freshwater genus, always occuring as two joined polyps, with a common stalk.

Phoronida is a small phylum of entirely marine animals consisting of sixteen species only; each one is equipped with a lophophore, or crown of feeding tentacles like the Ectoprocta. Their worm-like body is always housed inside a chitinous tube attached to rocks, shells or buried in sand. *Phoronis* [80, 81] and *Phoronopsis* are the two genera; the former have a narrow neck-like groove between the lophophore and the body, and the latter having an obvious collar-like ridge of tissue instead. Most species are solitary and reach up to 8 inches. A free swimming larval stage ensures dispersal.

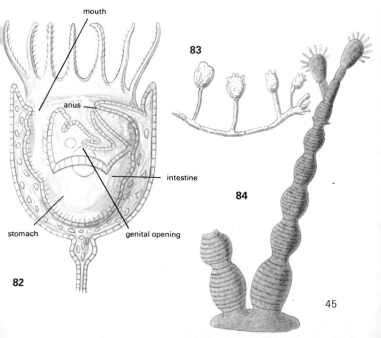

mouth

anus

intestine

stomach   genital opening

82

83

84

85

The phylum **Ectoprocta** contains fifty freshwater species, and many marine species.

Cristatella, Lophopus [85] and *Pectinatella* are freshwater species. They all have the individual animals, or zooids, embedded in a gelatinous colony sac, which in *Pectinatella* may form a flat patch up to 2 feet in diameter.

The marine species are some of the most numerous animals readily found on the seashore. Many species have developed two basic types of zooids. One type, the vibracula, are individuals which are mainly long bristles and serve the colony by creating a continuous current over its surface, keeping it free from detritus and bringing small planktonic organisms for food. The second type, the avicularia, are the feeding polyps and are armed with muscular beak-like jaws.

86

87

88

89

90

85 *Lophopus crystallinus*, a freshwater ectoproct
86 *Bugula*, a marine ectoproct
87 *Aetea*, a marine ectoproct
88 *Crisia*, a polymorphous ectoproct
89 *Stomatopora*, a polymorphous ectoproct
90 *Membranipora*, with shape of the cell case at high magnification

Bugula [86] is upright with branches that have zooids arranged in two rows up the stem. *Membranipora* [90] is an encrusting colony, with rectangular chambers, and is commonly found attached to the wracks (*Fucus* species).

Brachiopoda, the lamp shells, along with the phoronids and ectoprocts, are a true lophophorate phylum. They outwardly resemble the bivalve molluscs, but internally they have a lophophore, which is a complex ciliated feeding organ. The two shell valves are positioned dorsally and ventrally to enclose the soft parts, and the whole animal is attached by a pedicle, or stalk, either directly to the substrate, or at the base of a burrow.

There are only about 260 species of brachiopods living today; but in the past, they were a splendid and numerous phylum, reaching up to 30,000 species in the Palaeozoic and Mesozoic eras. The first record of their appearance in British rock strata, approximately 600 million years ago, is in the Cambrian shales of North Wales, where fossils of *Lingulella* exist. *Lingula* [92] itself dates back to the Ordovician period and probably holds the world record for longevity for any species. It is found in tropical waters, as are the majority of living species, for example, *Lacazella*, *Crania*, *Terebratula* and *Waldheimia*.

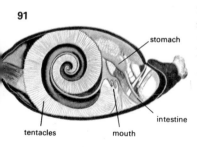

91

stomach

intestine

tentacles   mouth

91 Longitudinal section of
Magellania
92 *Lingula* in feeding position
within burrow

92

PHYLUM ANNELIDA

Three major phyla in the animal kingdom exhibit metameric segmentation, the annelids, the arthropods and the chordates. The ancestral annelids are thought to have evolved this pattern of similar body units by sheets of muscle developing between the bilaterally arranged viscera, mainly as a result of an undulatory method of locomotion. The arthropods are thought to rise from annelid stock and they retain this segmentation in the main. The chordates have evolved their metameric segmentation from an independent stock and it is first seen in the cephalochordates.

The metameres, or segments, are divisions of the body, arranged from head to tail in a linear series. They each contain similar structures in the perfect condition, such as muscles, excretory and reproductive organs, blood vessels and nerves, all arranged in like pattern. However, through the means of the nervous system, they are all co-ordinated into a whole unified body.

Superimposed on the metamerically segmented and coelomate animal is another basic trend, that is the tendency to develop a more complex anterior end, or cephalization. The head end of the annelids is more richly endowed with sense organs, feeding structures and consequently nervous co-ordinating structures, than any other part of the body. The arthropods show even greater development of the head. A similar pattern can be traced in the chordates, where the head becomes increasingly more specialized with respect to eyes, ears, nose and brain from fish to mammal.

As a phylum, the Annelida are identified by having a body wall consisting of a thin layer of cuticle overlying the epidermis, a layer of circular muscle immediately underneath, and an inner layer of longitudinal muscle fibres attached to connective tissue sheets. The nervous system consists of a pair of dorsal ganglia in front of the mouth, joined by two branches, around the gut to two ventral ganglia in the next segment. The rest of the segments have paired ventral ganglia joined longitudinally along the animal's length to form the ventral nerve cord. Each segment has branch nerves running to all its constituent parts. The digestive tract is a relatively simple tube from the mouth which lies behind the preoral segment, to the anus which is terminal. It is suspended in the coelomic cavity by a dorsal and ventral septum, and the lining cells secrete enzymes into its lumen to digest the food extracellularly. Excretion is performed by paired, segmentally arranged nephridia, which are open, ciliated funnels lying in the coelom, and leading to the exterior aperture, the nephridiopore, by a long coiled duct. The blood system is well developed, with a main dorsal vessel above the gut, in which the blood flows forwards, a ventral vessel below the gut, in which the flow is backwards, and interconnecting commissural vessels around the sides of the gut. Because of the thickness of the body wall and isolation of the gut by the coelom, there are also capillary networks in the skin and over the gut surface to help in distribution of oxygen and food products. This is known as a closed blood system as the circulating fluid is confined to vessels. In some leeches, this system has become modified.

There are three main annelid classes, the Polychaeta, including the marine bristle and paddle worms, the Oligochaeta, including the earthworms and some freshwater forms and the Hirudinea, the leeches, which are thought to have arisen from freshwater oligochaetes.

The Archiannelida, the Echiuroidea and the Siphunculoidea are given class status by some authorities and the latter two phylum status by others. But their affinities are obscure and the cause of minor controversy.

Archiannelida

This odd group of worms have lost most evidence of segmentation and have a ciliated epidermis. A few species possess bristles (setae or chaetae). Most species are marine, some that are able to tolerate a wide range of salinity have adopted salt pans as their habitat. They are minute, often brightly coloured, orange and red, and can be found clinging to vegetation or sifted from mud where they feed on the protozoan fauna. They are thought to be an artificial collection of aberrant polychaetes because the different genera do not show close affinities.

Sipunculoidea

The larval stages of these 'peanut worms' show three body segments, which rapidly disappear at maturity. This clue, along with the presence of a coelom and a single pair of nephridia, is the only apparent evidence of annelid affinity of this group. There are some 250 species, which range in distribution from the high water position on the shore down to the abyssal depths of the ocean floor. They range in size from a few millimetres to half a metre, and live in sand or mud, or take up tenancy in empty snail shells or worm tubes.

Echiuroidea

This third group of oddities has also lost segmental structure, and most species have lost all chaetae. They have developed a large preoral lobe which is extended over the surface of the substrate and forms a gutter along which food is sucked up. There are about seventy species known at present, but this number is increased as new species are discovered in deep sea dredges. More typical species live in mud and sand burrows, or in empty shells and sea urchin cases.

93 Medicinal Leech, *Hirudo medicinalis*; section showing crop pouches of left side distended with blood (fully fed) and right side empty
94 *Haementera costata*
95 *Glossiphonia*
96 *Haemopsis sanguisuga*
97 *Helobdella*
98 *Piscicola*

Hirudinea

These annelids have a fixed number of segments, thirty-three in total, which are confused externally by the presence of extra rings, or annuli, which do not continue inside the body as dividing partitions, or septa. They have no bristles and are more or less ovoid in cross-section, without the flap-like extensions of the body wall found in the polychaetes. Their main diagnostic feature is the presence of an anterior and a posterior sucker. They are all hermaphrodite, but most species cross-fertilize their eggs during copulation, when the sperm are exchanged. Some species, for example, *Glossiphonia* [95], brood their eggs by attaching cocoons to their ventral surface. The young attach themselves by their posterior sucker to the parent until they mature.

Most leeches are freshwater in habitat, but there are a few marine and terrestrial forms; in total there are over 300 different species. Not all leeches are parasitic; some are scavengers; others are predators. The terrestrial European leech, *Haemopsis sanguisuga* [96], can live some way from water and frequently swallows whole earthworms! Blood sucking leeches attach a variety of hosts, some are specifically snail leeches, for example *Helobdella* [97], or fish leeches, *Piscicola* [98]. Others like the Horse Leech, *Haemopsis marmorata*, live in ponds and troughs and will take a meal off a variety of livestock. The Medicinal Leech, *Hirudo medicinalis* [93], used to be used for 'blood letting' until this was stopped as a medicinal practice. Most leeches are small, harmless animals, about 2 to 3 centimetres in length, and are often mottled in green, red and brown colours.

101

Polychaeta

The polychaetes conform well to the basic annelid plan, and are typified by the presence of parapodia, which are lobed extensions of the lateral body wall, mainly bearing numerous chaetae. The head end is always well defined, and often has tentacles and sometimes eyes. They are entirely marine and are divided into two major sub-groups. The errant, or free moving forms, range from pelagic species such as *Tomopteris* [103], through surface crawlers like the scale worms, to the active swimmers which hunt for food but may dwell in simple mucus-lined burrows when not feeding, such as *Nereis*. They also include active burrowing forms which tend to lose their chaetae and head appendages. A few families consist of ectoparasites, and one family consists of extremely modified commensal and parasitic forms, for example, *Myzostoma*, which is like a flat disc of tissue with a fringe of cirri.

The sedentary polychaetes are typically tube dwellers, and because of their less active life, their parapodia have undergone various stages of reduction or modification. Very often, the head end becomes elaborated, and the rest of the body reduced as in the fan worms. *Chaetopterus* [105], has developed its parapodia for many different functions as it lies in its U-shaped burrow; some fan the water and maintain a feeding current, two others secrete and hold a mucus bag in which it collects food, a ciliated cup catches up the entangled food mass and passes it forwards to a ciliated tract which leads to the mouth – a very complex series of modifications. The sedentary, tubicolous habit has meant that many of these polychaetes have sacrificed vital surface area for oxygen absorption, so many species have developed filamentous gills, either in irregularly branched tufts like the lugworms, or as long simple threads as in *Cirratulus*. The tubes are mainly constructed of sand grains, cemented together with mucus. The tube of *Chaetopterus* [105] is parchment like, others are more membranous.

The polychaetes feed in a whole range of ways and on all types of food from detritus, protozoa and small algae to whole soft-bodied invertebrates, even other related species. The sedentary polychaetes are nearly all ciliary feeders; they trap suspended food particles in mucus on their tentacles, and the cilia waft the laiden mucus to the mouth.

Most polychaetes are dioecious, that is unisexual, and several families become pelargic at sexual maturity, so that they all swarm to the surface layers and shed the eggs and sperm simultaneously into the water. The adults die after the gametes have ruptured from the gonads.

99 *Sabella*, a tubicolous polychaete
100 *Aphrodite* an errant polychaete
101 *Flabelligera*
102 *Amphitrite*, tubicolous polychaete
103 *Tomopteris*
104 *Lepidonotus*, errant polychaete
105 *Chaetopterus*

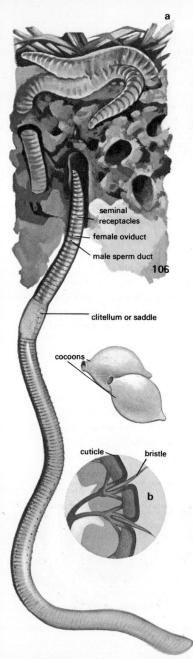

seminal
receptacles

female oviduct

male sperm duct

106

clitellum or saddle

cocoons

cuticle   bristle

b

Oligochaeta

The oligochaetes are a less varied group than the marine polychaetes. They do not have parapodia, so that the body is basically cylindrical in cross section, and chaetae (bristles) arise singly out of a muscular sac, rather than in numbers as in the polychaetes. The head end is reduced in comparison to the polychaetes, because of the burrowing habits of the majority of species; there are no obvious eyes, except in the few freshwater species with light-sensitive cells, which prevent them exposing themselves to damaging light.

The oligochaetes are all hermaphrodite; the gonads are not repeated in segments down the body, but are restricted to a few definite segments at the anterior end of the body. Copulation ensures cross fertilization of the eggs, which are laid in lemon-shaped cocoons [106]. The eggs provide the developing worms with sufficient food to enable them to hatch out in an advanced state.

106 *Lumbricus terrestris,* the Earthworm a. copulating b. detail of chaetae
107 *Tubifex: a.* in tubes
108 *Stephanodrilus*

There are about 3,000 species of oligochaetes, which include freshwater and terrestrial types ranging in size from 1 millimetre to 12 feet, as seen in the giant tropical earthworm. The largest of all is the Australian species, *Megascolides australis*. The bottoms of ponds, lakes, streams and rivers, particularly muddy and silty conditions, are ideal spots to find aquatic oligochaetes. Terrestrial oligochaetes, or earthworms, are world-wide in range with the exception of the polar ice-caps, and regions of perma-frost.

The freshwater forms sometimes resemble the poly-chaetes in that they have gills and tufts of chaetae. *Tubifex* [107] is probably the best known freshwater genus is sold in pet shops for amphibian and fish food. Its natural habitat is mud, where it sits head down in its burrow [107a].

The earthworms live in burrows which are lined with mucus-bound soil particles. The burrows can be as deep as 8 feet, penetrating well into the subsoil layer if there are particularly dry or cold climatic conditions. The worms coil together in knots, and encase themselves in a cocoon of soil particles and mucus in time of adversity.

They are mainly active at night when they come to the surface and collect leaves and other organic detritus, dragging it into the burrow to be consumed. The presence of earth-worms in the soil has long been acknowledged as invaluable to the formation of fertile topsoil. Recent estimates quote figures of half-a-million to two-and-a-half-million earth-worms per acre for arable and pasture land. They help with aeration and drainage of the soil, bringing up subsoil to add to the topsoil, at about 1 inch every five years. They also neutralize and increase the organic manurial content.

109   109 Spider

110 Centipedes

110

PHYLUM ARTHROPODA

Eighty per cent of all recorded living species of animals are arthropods, representing some 800,000 or more types and millions more actual individuals. They have successfully conquered all environments, and are the only invertebrate group to have mastered the art of flight.

The arthropods are thought to have arisen from early annelid stock and they still resemble that group in that they retain clear metameric segmentation in all the embryonic stages. Most adults can be seen to be segmented, but many forms, such as spiders, ticks and mites, lose the segmental pattern in the adult. They also have, in the primitive condition, one pair of jointed appendages per body segment, though this becomes modified in many classes. There is always one pair of the appendages which acts as jaws. The nervous system is similar to the annelids with a double ventral nerve cord swelling into ganglia in each segment.

The most important evolutionary change from annelid to arthropod has been the development of a hard exoskeleton, an outer covering of cuticle, which in some cases is extremely thickened by the addition of calcium salts. The presence of such an exoskeleton has meant that the arthropods can develop tough limbs for a variety of purposes, for example, the grasping claws of crabs and lobsters, the tail barbs of scorpions, the piercing mouthparts of many insects. The cuticle has to be shed to allow for growth. After shedding, or ecdysis, the new cuticle is soft and permits body expansion, but it soon hardens and prevents growth until the next shedding. The main disadvantage of ecdysis is that when the cuticle is new and soft the animal is vulnerable and easy prey to attackers [111].

Correlated with the presence of a cuticle is the problem of breathing, and the arthropods have developed a wide variety of gills, trachea, lung books, and other respiratory structures which enable oxygen to be readily absorbed into the blood spaces.

The classification of the arthropods is controversial. The main living classes are the Onychophora, the Crustacea, the Myriapoda, the Insecta and the Arachnida.

111

111 A dragonfly; a. adult, b. emergence, c. nymph

a

b

c

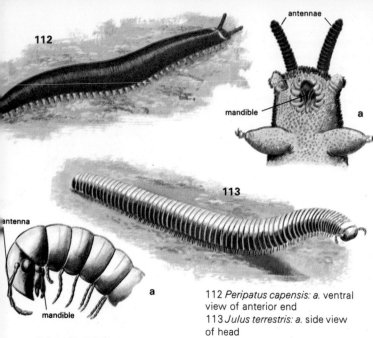

112 *Peripatus capensis: a.* ventral view of anterior end
113 *Julus terrestris: a.* side view of head

Onychophora

These arthropods are an ancient group, which from fossil evidence do not appear to have changed much from the Cambrian period. There are about seventy living species of onychophorans and they inhabit two distinct climatic regions: the tropics, where only members of the family Peripatopsidae are found; and the southern temperate belt, where only the family Peripatidae are found.

The onychophorans are mainly found in very humid conditions because they have not acquired the waterproof cuticle of later arthropods. During periods of drought or low temperatures, they hide away in protective burrows deep in leaf litter. They are rather slug-like, with short stumpy legs along the body, each terminating in a pair of small claws. The head bears two ringed antennae, a pair of secretory lobes and a ventrally situated mouth armed with chitinous jaws, the mandibles. They are omnivorous, feeding on vegetation, snails, small insects and worms, several species seem partial to termites.

Myriapoda

The elongated body with many leg-bearing segments was the main feature considered when uniting the centipedes and millipedes and two minor groups, the pauropods and the symphylans, into this one class, the Myriapoda.

The centipedes, or Chilopoda, are flattened dorso-ventrally, with one pair of antennae, three pairs of jaws, and a pair of segmented appendages per body segment, except the last two. The first pair immediately behind the jaws are claws which house a poison gland. The rest are simply walking legs [110]. There are probably about 5,000 species of centipedes in the world, distributed in temperate and tropical regions and readily found in soil, leaf litter, and under logs and stones along with woodlice, slugs and insects.

During the development of the millipedes, or Diplopoda, two body segments have become fused together so producing double or 'diplo'-segments, each equipped with two pairs of legs. They have two pairs of jaws in the adult. The tropical millipedes reach about 28 centimetres in length, and the most common forms are the bluish-black, cylindrical *Julus* [113]. Being slow movers, many have developed toxic and repellant secretions for defence.

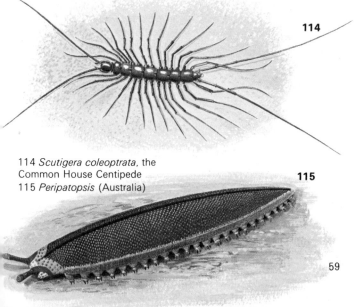

114

115

114 *Scutigera coleoptrata*, the Common House Centipede
115 *Peripatopsis* (Australia)

116

117

a   b

118   **120**

119

Crustacea

The crustaceans are distinguished as a class by the presence of two pairs of antennae as the most anterior head appendages of the body. Next are the mandibles, followed by two pairs of ancillary feeding structures, the first and second maxillae. The appendages of the trunk are less uniform. The enormous variation of body, limb presence and function in the crustaceans results in about 26,000 species of marine, freshwater and a few terrestrial animals.

The fairy shrimps, tadpole shrimps, water fleas and clam shrimps all form the subclass **Branchiopoda**, whose trunk appendages are flattened, leaf-like, and function as gills. The tadpole shrimps have a large crescentic cara-

116 Goose Barnacles, *Lepas anatifera*
117 Acorn Barnacles, *Balanus balanoides*. The free-swimming stages, *a*. nauplius *b*. cypris, clearly show their affinity to other crustaceans.
118 *Elimnius morestus* has four plates and was introduced into southern England from Australia. It is rapidly becoming common.
119 *Balanus perforatus*, the largest of the barnacles; may be $1\frac{1}{4}$ ins tall
120 *Balanus crenatus*

pace, rather resembling the king crab's, but much smaller. One of the fairy shrimps, the Brine Shrimp, *Artemia,* is found in salt lakes. The water fleas, such as *Podon* and *Apus,* have respresentatives in sea and inland waters; *Daphnia* [124] is the most common freshwater genus.

The **Ostracoda** have the least number of appendages of all crustaceans, and the whole body is enclosed by a bivalved carapace. Most species are minute and freshwater dwelling; but there are several marine genera, and one deep water form, *Gigantocypris*. One or two terrestrial ostracods have also been recorded.

The **Copepoda** are the most numerous in species and numbers of all crustacea and about ninety per cent of most plankton hauls will be copepods such as *Calanus* [121]. Most of the 4,500 species are marine, a few are freshwater, and over 1,000 are parasitic, such as *Caligus* an ectoparasite on fishes' gills.

The **Cirripedia** are more commonly known as the barnacles, although there are a few parasitic species not included within this common name. The larval cirripedes are free-swimming, but they settle down on a suitable rock, shell, wood or even metal surface and develop calcareous plates to form the typical armoured barnacle. *Lepas,* the Goose Barnacle [116], and *Scalpellum* are both off-shore stalked barnacles. *Sacculina* is parasitic on crabs.

The sub-class of crustaceans **Malacostraca**, includes the more familiar forms such as crabs, lobsters, shrimps, woodlice and sand hoppers.

The order Isopoda is the second largest group containing some 4,000 species all with the common feature

121 *Calanus*   122 *Podon*   123 *Apus*   124 *Daphnia*

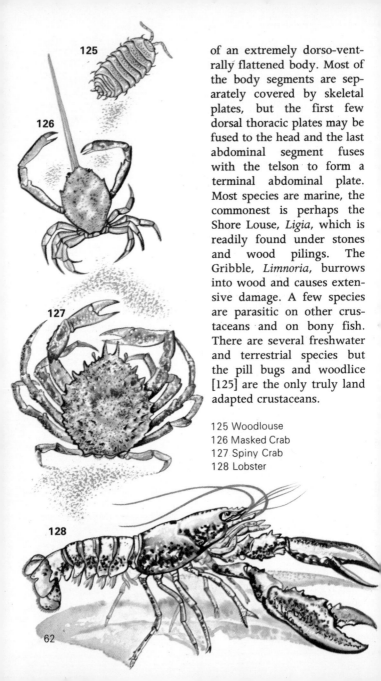

of an extremely dorso-ventrally flattened body. Most of the body segments are separately covered by skeletal plates, but the first few dorsal thoracic plates may be fused to the head and the last abdominal segment fuses with the telson to form a terminal abdominal plate. Most species are marine, the commonest is perhaps the Shore Louse, *Ligia*, which is readily found under stones and wood pilings. The Gribble, *Limnoria*, burrows into wood and causes extensive damage. A few species are parasitic on other crustaceans and on bony fish. There are several freshwater and terrestrial species but the pill bugs and woodlice [125] are the only truly land adapted crustaceans.

125 Woodlouse
126 Masked Crab
127 Spiny Crab
128 Lobster

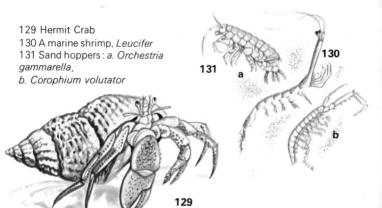

129 Hermit Crab
130 A marine shrimp, *Leucifer*
131 Sand hoppers : *a. Orchestria gammarella,*
b. Corophium volutator

131

130

a

b

129

The Amphipoda are rather similar to the Isopoda in general bony structure, with the clear distinction of a laterally flattened body, rather than a dorso-ventrally flattened one. They live mainly in marine conditions [130], but there are some freshwater and fewer semi-terrestrial forms. They have mastered a wide range of habitats, from thick mud burrowing, wood boring and sand tunnelling, through to the less industrious habits of clinging to seaweeds, and of free-floating as part of the plankton. *Gammarus* commonly called the Freshwater Shrimp is a freshwater genus used frequently in the laboratory.

There are several small orders of crustacea which are beyond the scope of this book, but worthy of brief mention are the Euphausid shrimps, which occur in swarms in plankton and form the main food for whales, and are filtered off literally by the ton.

The Decapoda are the largest group of the Crustacea, containing some 8,500 known species, mainly marine. Crayfish, a few true shrimps, and a few crabs are found in freshwater and even fewer crabs are lake-shore and land dwelling. The group is distinguished by the first three pairs of thoracic appendages becoming auxiliary mouth parts – the maxillipeds; and the fourth to eighth pairs forming ten walking legs, hence Decapoda. The first pair of walking legs is usually armed with heavy pincers, or chelae, which are used for feeding and defence as in the Lobster [128].

132 Life cycle of the Swallowtail; *a.* adult, *b.* egg (laid singly), *c.* young larva, *d.* caterpillar, *e.* pupa

Insecta

The class Insecta is the most numerous class of animals in the whole kingdom. There are several thousand new species described each year, and the suggested final numbers of species range between four and five million.

The insect body is generally divided into three distinct regions, head, thorax and abdomen. The head is made up of six fused body segments and bears one pair of antennae, a pair of compound eyes, which may be reduced or absent in some species, and mouthparts adapted for different types of eating. The thorax consists of three segments, each bearing a pair of jointed legs. The abdomen has twelve segments except in the primitive springtails, which have six; the last segments are equipped with appendages which serve as genital

133 Periodical Cicada; *a.* egg and egg scars, *b.* nymph, *c.* emerging adult, *d.* adult Cicada

organs. Insects breathe with a system of internal tubules, the trachea, which open by specialized paired apertures down the body called spiracles.

The body wall of insects has an outer layer of cuticle which is part chitinized; the surface is waterproof, protective and sometimes merely for ornament. The smallest insects are less than $\frac{1}{4}$ millimetre long, the largest include the stick [136] and leaf insects [135] some 30 centimetres long.

The life histories of insects follow two basic patterns. The more primitive insects lay eggs which hatch into a larva which becomes an active nymph. The nymph resembles the adult except that it lacks wings. It becomes progressively more and more like the adult, or imago, by a series of moults, each recognizable stage being called an instar. This pattern of development is termed incomplete metamorphosis [133]. The higher insect orders hatch into a larva or grub, which has a resting stage after feeding, known as the pupa. In this state, the insect undergoes complete reorganization of tissues, emerging as the imago. This is complete metamorphosis [132].

The mouthparts of insects are adapted to many specialized modes of feeding [134]. The simplest arrangement is the chewing mouthparts of the primitive insects, such as the locusts and dragonflies. The worker bee has maxillae modified for licking and sucking nectar. The butterflies and moths have a long coiled proboscis and have lost the mandibles completely. The insects which pierce animals or plant tissues usually have mandibles and maxillae.

134 Forms of mouthparts : *a.* butterfly, *b.* bee, *c.* mosquito, *d.* locust

135

136

Insects show various degrees of social groupings from solitary animals, to those which are gregarious due to environmental factors, and to those which are true societies.

The migratory locust passes through a series of generations as a solitary animal, then due to factors which are not properly understood, they become the migratory, swarming locusts that cause such devastation. Many butterflies and moths are gregarious from time to time, creating migrating flocks; the Cotton Moth is a serious cotton pest and is an active migrant. Larvae in various orders congregate and march in armies causing devastation.

The true social insects, such as the wasps, hornets, Bumble Bees, honey bees and ants, live in communities all the time. There are castes in all species, for example, in the honey bees [138] there are workers, drones and queens. The Paper Wasps [141] build a nest of chewed wood and saliva, and the Bumble Bees have subterranean nests built of earth, wax and resin [142]. Social life in the ants is highly

135 *Phyllium*, Leaf Insect
136 Stick Insect, *Carausius morosus*
137 African termites : *a.* termite hill, *b.* male, *c.* queen, *d.* soldier, *e.* worker

137

evolved; some species live in rotting wood, others produce galls in living trees, but the majority excavate underground nests with galleries, breeding chambers and storage spaces. Termites build nests which range in size from small wood-pulp mounds to the 20 feet high earth constructions of *Bellicostermes* near Lake Rudolph [137].

Insects have many defence mechanisms. They may have stings or poisonous bites and frequently produce obnoxious secretions which ward off enemies. If caught, some species will shed limbs to escape and can regenerate them. Cryptic coloration is the condition where the insects are perfectly matched to their normal resting background. Protective resemblance is seen in the stick and leaf insects [135, 136]. Mimicry is the phenomenon of one vulnerable species looking like invulnerable animals of the same habitat.

138 Honey bees: *a.* worker,
b. drone, *c.* queen, *d.* queen cell
139 *Dolichorespula saxonia*,
Saxon Wasp
140 Potter Wasp
141 *Polistes bimaculatus*,
Paper Wasp
142 *Bombus terrestris*,
Bumble Bee

Many insects are adapted for a parasitic way of life, obtaining their nourishment at the expense of another animal. In turn, other animals use the parasites as carriers or vectors, and in this respect, the parasitic insects are most dangerous to Man and his domestic animals.

Lice live on a range of animals, clinging to the skin and feeding on blood. Birds and mammals are particularly subject to louse infection. Man is the natural host of the Pubic Louse and Body and Head Louse [145]. The latter is a carrier of many diseases, including typhus.

The fleas are also blood-sucking parasites, but can exist for long periods without the host species in larval and adult form. The Dog Flea is the vector of common dog and cat tapeworms. Rat Fleas [144] are vectors of bubonic plague. The bed bugs [146] are a small family of avian and mammalian ectoparasites prolific in slum conditions.

143 Housefly: *a.* adult, *b.* eggs, *c.* maggot, *d.* pupa
144 Rat Flea, carrier of bubonic plague
145 Body Louse, carrier of typhus and other diseases
146 Bed bug

The Diptera, the flies and mosquitoes, include vectors of some of the most dangerous and crippling diseases of man. Yellow fever and malaria are transmitted by various mosquitoes, as is elephantiasis, dengue fever, and encephalitis.

The houseflies [143] are serious transporters of bacteria in the average household. When they feed, they vomit and excrete on the food, and are known vectors of some forty different diseases, including typhoid, tuberculosis and amoebic dysentery. The tsetse flies, *Glossina* species, are notorious vectors of sleeping sickness, the trypanosome infection which affects Man in Africa; and of n'gana which affects cattle. Diseases of Man and his animals apart, the insects take their toll of crops, cereals, plantations, stored food and timber.

Locusts [148] devastate whole regions, completely devouring every living green leaf. Leaf-hoppers damage garden, orchard and vineyard, pasture land and cereal crops. They cause serious economic loss to rice plantations, introduce other diseases and stunt growth of plants by feeding on the juices. Stink Bugs [149] also feed on plant juices and caterpillars [150] feed on the plants themselves.

Aphids [147] include greenflies, blackflies and plant lice. They occur in enormous numbers as the females can reproduce parthenogenetically throughout the summer, the fecundity is hardly rivalled in the animal kingdom. They basically suck the plant sap and cause it to wither and die.

147 Aphids   148 Locust   149 Stink Bug   150 A caterpillar

Arachnida

The arachnids differ from the insect arthropods in that the head and thorax are fused to give a single body unit, the prosoma, and the posterior part of the body, consisting of thirteen segments and a telson, forming the opisthosoma. The appendages on the prosoma are one pair of prehensile pincers, one pair of pedipalps, either sensory or prehensile in function, and four pairs of walking legs.

The scorpions are the oldest known terrestrial arachnids with fossil records dating from the Silurian. The majority of species are humid forest and jungle species, hiding by day under leaf litter and hunting at night. Fewer species inhabit the arid desert regions. The sting of scorpions is attached to the last segment and the curved barb injects the venom, secreted by two glands.

Many species are not deadly toxic to man, and use their

151 Sheep Tick, *Ixodes ricinus*
152 Water Mite, *Hygrobates longipalpis*
153 *Buthus,* the Scorpion
154 Sea Spider
155 King Crab, *Limulus polyphemus*
156 Orb Spider
157 Bird-eating Spider

156

157

sting to capture invertebrate prey. The African genus *Androctonus* and some species of centruroides are notoriously venomous. *Buthus* [153] is a typical scorpion, bearing its young alive and preparing for copulation with an intricate courtship dance by the mating pair. The King or Horseshoe Crabs [155] are aquatic arachnids, some taxonomists place them in a separate class, the Merostomata. *Limulus* is the only representative found on the northwest coast of America. The other two genera are found on the Asian coast.

The largest order of Arachnida is the Aranae, the spiders, comprising some 30,000 species. They are distinctive in their production of silk threads used for web spinning [156], and for protecting the eggs. The pedipalps in the male are adapted to form copulatory organs. The first pair of appendages, the chelicerae, are pointed, fang-like, and are provided with poison glands. A few species of the Black Widow spiders are neurotoxic to Man, but fatal cases are usually restricted to small children. Some female spiders are infamous for their habit of devouring their well-meaning mates and several species have overcome this problem by developing males smaller than the normal food animals, which do not trigger off the pouncing action of the female.

The mites [152] and ticks [151], though including many free-living aquatic and terrestrial forms, also include some of man's most hated pests. Many species are bloodsucking parasites on Man himself, while others attack domestic animals, crops, stored cereals and many other products.

Sea spiders [154], the Pycnogonids, form a small group of curious arachnids. Most species are small, ranging from 1 to 12 millimetres. *Pycnogonium* is a common seashore spider.

158 *Neopilina: a.* ventral view,
b. cross-section
159 Chiton : *a.* ventral view,
b. dorsal view, *c.* shell plate and
marginal girdle enlarged
160 Four kinds of tusk shell

PHYLUM MOLLUSCA

The molluscs are the second largest invertebrate phylum, containing over 80,000 described living species. Because of the nature of their shells, which are easily fossilized, they rank among the most important and numerous fossils; over 35,000 fossil forms are known and used as geological indices, relating strata in one part of the world to another, thus helping to establish the correct sequence and time scale.

Molluscs have a solid, unsegmented body which can be divided into four main parts unless greatly modified. First is the head, bearing tentacles, eyes and other sense organs. The mouth opens close to the anterior end and is often armed with teeth. Internally lies the brain, or cerebral ganglia, which is highly developed in the cephalopods. The visceral sac is the next distinct part which houses the gut, heart and reproductive organs. The foot is a muscular structure used for locomotion and burrowing and is developed into tentacles in the cephalopods. The mantle is the fourth region of the body and is a fold of skin which develops from the posterior part and folds over, enveloping the visceral mass and separating it from the shell, thus forming a cavity between the shell and the rest of the body. The anus and excretory ducts open into the mantle cavity inside which the gills develop. The shell is secreted by the edge of the mantle and consists of an outer horny layer, with an internal pearly or nacreous layer of calcium carbonate.

The molluscs as a group show the widest possible adaptive radiation, they have mastered all marine, freshwater and terrestrial habitats from the abyss to deserts. The only thing they cannot do is fly which the Arthropoda have managed.

The Amphibeura are the most primitive molluscs without eyes or tentacles and with a dorsal shell, sometimes composed of separate plates. The chitons [159], or coat of mail shells, belong to this group. They are common on the seashore but are small and well camouflaged, so may not always be noticed. The giant Pacific species, *Amicula stellari*, is over a foot in length, and the shell is overgrown by the soft mantle tissue.

Neopilina [158] is a recent zoological specimen; it was found in dredgings off the Costa Rican coast in 1952 and resembles some Cambrian molluscan fossils. It has a primitive gill array and is thought to be a link between the segmented annelids and un-segmented molluscs.

The Scaphopoda, or Elephant's Tusk Shells [160], have tubular shells open at both ends. They have a reduced foot and no gills. *Dentalium* has an oval ring of tentacles which catch food with their sucker-like ends.

160

161

162

Gastropoda

This class is the largest molluscan group containing over 35,000 species of living gastropods and some 15,000 fossil forms. Marine, freshwater and land forms live in all types of conditions. The conquest of land has been made possible by the changeover of the gills in the mantle cavity to a vascular lung, operating as any terrestrial lung surface.

The larval gastropod undergoes a twisting or torsion during its development, which results in the internal viscera being displaced. The visceral mass may also be coiled to a greater or lesser extent and the degree of coiling is reflected in the coiling of the shell.

The limpets show little or no coiling of the shell. *Patella*, the Common Shore Limpet, shows none. *Haliotis*, the Abalone or Ormer [161], retains a flattened spire at the end of its ear-shaped shell. *Diodora*, the Keyhole Limpet, and *Lacuna*, the Emperor's Chink Shell, both have shell apertures as exits for the gill current. The coiled spires of the winkles sting winkles, top shells, whelks and hundreds of large tropical shells are commonly known. The gastropods exhibit every type of ornamentation of the spire and aperture, coloration and pattern. *Lambis*, the Spider Shell [163], has an aperture which is drawn out into curved projections.

164

163

Conus [165] shells have a reduced spire and large last whorl to house the body.

Gastropods employ a whole range of feeding habits. The radula, a constantly growing rasping tongue, is used by vegetarian species to rasp away at rock surfaces to remove the algal growths. The land snails eat vegetation in a similar way. The *Murex* shells [162, 164], sting winkles and whelks are carnivorous, feeding almost exclusively on bivalve molluscs The cone shells stab their prey, with their poisonous radula teeth, then grind them up at leisure. Some ciliary feeding gastropods, like the Slipper Limpet, and *Vermetus*, an uncoiling genus, trap plankton on the gill surface which is covered in mucus, and then eat the mucus string by passing it to the mouth.

161 *Haliotis*
162 *Murex*
163 *Lambis*
164 *Murex regius*
165 *Conus,* Cone Shell
166 *Epitonium*
167 *Ficus,* Fig Shell

Opisthobranchia and Pulmonata

The Opisthobranchia are gastropods which undergo uncoiling of the visceral mass and gradual reduction or loss of shell.

The tectibrachs still have a shell, but it is often hidden in the folds of the extended mantle. The bubble shells, *Actaeon, Akera* and *Cylichna*, show a graded reduction of the shell from a spired form with a large aperture to a simple cylinder without ends. The nudibranchs have little or no shell and the original gills are lost; the dorsal surface of the body is extended into cerata or respiratory tentacles. These forms are the sea slugs [168, 169], which include some of the most colourful and aesthetic marine specimens. *Aplysia*, the Sea Hare [170], is easily found at low tide when it comes inshore to lay its eggs. Normally it lives below the tide limit and browses on the large brown algae. The flap-like extensions of the body are used for swimming actively, normally it

168 Sea Clown
169 Orange Marcinella
170 Sea Hare

171 Giant Land Snail with eggs
172 Slug
173 Common Snail and eggs

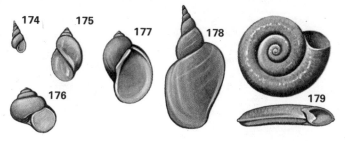

crawls on its foot. The paper-thin shell remnant is hidden by the mantle on its back. *Facelina* has white-tipped gill tufts down the sides of its cerise to pale pink body. *Dendronotus* is perfectly camouflaged in brown with reddish flecks.

The Pulmonata are gastropods which use the mantle cavity as a lung, and include the freshwater and terrestrial species. The freshwater limpet, *Ancylus*, may be found in lakes and streams. Its eggs are laid in vast quantities of albumen and surrounded by an egg case which is attached to vegetation or the substratum. *Limnacaea* is the genus of spired pond snails [177, 178]; the shallow water forms visit the surface at intervals and replenish the supply of air to the mantle cavity. Deep water forms, such as *Limnaea abyssalis,* keeps the cavity full of water. The planobids have a flat coiled shell, like a coiled table mat, and the hydrobias are small, blunt spired forms, commonly found in muddy deposits. The terrestrial snails are typified by *Helis,* which is herbivorous. *H. aspersa* is the Common British Snail [173].

The slugs have lost their shell in most species and most are herbivorous [172]. The pteropods are pelagic opistho-branchs, called the sea butterflies [180], lightweight, delicate forms which spend their life in the plankton.

174 Jenkin's Spire Shell
175 Bladder Shell
176 Valve Shell
177 *Limnaea ovata*
178 Great Pond Snail
179 Ramshorn
180 Sea butterfly
181 *Tethys leporina*, a nudibranch

Bivalvia

As the name implies, the bivalves, correctly termed the Lamellibranchiata or Pelecypoda, have two halves or valves to the shell which enclose the soft parts of the body. Both valves can be closed by adductor muscles which pull from valve to valve on the inside. The elastic ligament at the hinge forces the valves to spring apart when the muscles relax. The hinge is the dorsal junction of the valves and may be intricately toothed and notched. The head is very much reduced and lacks the eyes, tentacles and toothed radula of the gastropods. The mantle cavity is spacious and in it lie the large sheet-like ciliated gills which act as a respiratory surface and as feeding organs for these animals which are filter feeders. The foot is used in many forms for burrowing into sand and mud. In attached forms, the foot is reduced or absent. In burrowers, the mantle edge becomes extended to form tube-like siphons through which the respiratory and feeding current is maintained [182].

The majority of this group are burrowers in sand and mud, or anchored to the surface. A few species are specialized borers; some can swim actively; a few forms become parasites. The sizes range from the freshwater seed shells of 2 millimetres length to the giant south Pacific clams, such as *Tridacna* [184], which reaches 4 feet in length and weighs over 500 pounds, flesh and shell.

Nucula is a primitive bivalve that burrows under the surface sand. Cockles and venus shells have short siphons and do not burrow deeply. Cockles [183] make progress over the surface of the sand by a kicking action of the foot. The razor shells, *Ensis* species, are suited to burrowing with the knife-like shells and muscular foot. The gapers cannot close their shell completely and have a thick leathery sheath around the siphons for protection. *Mya* is a mud dwelling gaper [189], the piddocks are specialized for boring into harder substrate and *Pholas* [188] and *Barnea* have specially ridged shells which rasp away rock.

184 *Tridacna*, Giant Clam
185 *Hippopus*, Bear's Paw
186 *Ensis arcuata*
187 *Ensis siliqua*
188 Common Piddock, *Pholas dactylus*
189 Blunt Gaper, *Mya truncata*

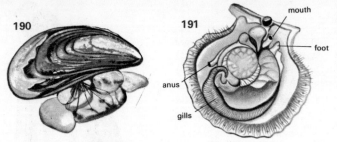

190

191

mouth

foot

anus

gills

The Common North Atlantic Shipworm, *Teredo navalis* [194], has greatly elongated siphons and small valves at the anterior end. The valves rock sideways on the rounded hinge and as the adductor muscles relax and contract, the valves' razor-like edges chisel away at the wood. The animal turns direction in boring to produce a U-shaped burrow. The siphon sheath secretes a calcareous lining on to the wood. Considerable damage is done each year to jetties and pilings throughout the world by this destructive bivalve.

The mussels anchor themselves by means of horny threads, the byssus, which are secreted by a gland in the foot [190]. They are found attached in clusters to rocks, gravel banks and wharfs. The Saddle Oyster, *Anomia*, is attached by a byssus which extends through a hole in the right valve.

The pearl oysters and scallops [191] also lie on the right valve, with several possible results; both valves may become concave or the left one may become flat as in the Queen Scallop. If a foreign particle becomes lodged between the mantle and the shell, it becomes coated by the inner most nacreous layer of the shell and forms a pearl [192]. Most oysters form pearls in this way but the finest, most expensive

193

192

commercial pearls are grown in *Pinctada* species of warm Pacific waters.

Anodonta [193] and *Unio* are freshwater mussels which drag themselves slowly across silt by foot action. The larvae of these freshwater forms have a fascinating parasitic existence. The eggs are held at first by the adult in the gill chamber. When they reach the larval stage, they have two small valves with clusters of sensory bristles at the edge and hooks in the case of *Anodonta*. They sink to the bottom of the water and wait until they come in contact with a fish. Then they clamp the valves closed over the fins, or scales, or gills and become imbedded by scar tissue that the fish grows in response to their presence. They live on the host's tissues until more adult features develop and then break free to live an independent existence.

190 Common Mussel attached to small stones by byssal threads
191 Anatomy of a scallop
192 Pearl Mussel, *Margaritifer margaritifera*
193 Swan Mussel, *Anodonta cygnea*
194 *Teredo*, Shipworm
195 *Lithophagus*, Date Mussels which bore by acid secretion

194

195

196 Giant Squid
197 Deep sea squid with
luminous organs

Cephalopoda

The most striking feature of the cephalopods is the well developed head, bearing a pair of eyes and a crown of tentacles which are derived from the foot. The tentacles are extremely prehensile and have their inner surface lined with sucker pads. The octopuses have eight equal length arms, and the squids have eight short and two long arms, ten in all, giving rise to the names Octopoda and Decapoda respectively. The class as a whole is well adapted for swimming. The mantle cavity of all cephalopods can be filled with water by expanding the cavity then by rapid contraction they expel the water and propel themselves backwards.

The **Decapoda** have an internal shell; the cuttle bones are the spongy shells washed up on the shore after the animal dies. They are all

198

carnivorous and have a pair of powerful beak-like jaws as well as a radula [199]. *Loligo,* the Common Squid, bites at the neck of the fish, severs the nerve cord and then the two long tentacular arms wrap up the prey and hold it to the mouth for further consumption. An ink sac is produced in an intestinal gland near the anus. When alarmed, it releases the ink to form a dense cloud behind which it can rapidly retreat. The ink of *Sepia* [198], the Cuttlefish, is of commercial use.

The dermis of squids contains many chromatophores, more are distributed in the dorsal than the ventral surface. They consist of a central bag of pigment surrounded by radiating muscle fibres [199]. As the fibres contract, the elastic wall of the sac is pulled outwards to give an enlarged pigment area, thus deepening the colour. *Sepia* [198] is capable of a phenomenal range of colour and patterning.

Loligo and *Sepia* are found close to the shore. *Architeuthis,* the Giant Squid [196] of the North Atlantic is said to reach 60 feet in length, though most of this is tentacles. Deep sea

199 Anatomy of a squid with detail of cells

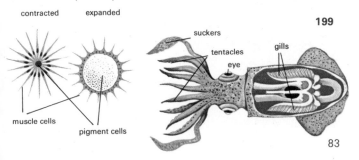

contracted　expanded

199

muscle cells

pigment cells

suckers

tentacles

eye

gills

squids [197] have light organs scattered over the body. Most of them emit a pale bluish-green light, but *Lycoteuthis* has a layer of red pigment and emits a rich ruby light.

The **Octopoda** have lost all traces of shell in living forms, the last genus to retain a small shell was *Palaeoctopus*. *Octopus* and *Eledone* are typical inshore forms, with a bag-like body, no fins and eight long arms [202]. The arms are extremely sensitive and help to locate food as well as trap it. *Eledonella* is a deep sea octopod that has been taken from depths up to 17,000 feet. *Amphitretus* has webs between the arms which may help prevent sinking, acting like a parachute, and are used to swim with, rather than proceeding by jet propulsion like the other cephalopods. *Cirroteuthis* and *Opisthoteuthis* are also web swimmers. *Cirroteuthis* has a conical body and swims by opening and closing the web gently. *Opisthoteuthis* is flattened and holds the mouth downwards and the arms spread outwards so that it looks like an eight armed starfish.

The argonauts are found in the upper waters of the open

201

200 Ammonites
201 *Nautilus*
202 Octopus with eggs

sea from the surface to depths of 3,000 feet. *Argonauta* is particularly noteworthy in that the sexes are very different from each other. The male is shell-less and small, but typically octopoid in appearance, whereas the female develops a thin pearly shell which is secreted by two modified arm folds. This shell forms a boat in which bunches of eggs are brooded. *Ocythoe* is a pelagic, shell-less argonaut where the male is dwarfed even more than in *Argonauta*. It is only $\frac{1}{2}$ inch long and shelters inside tunicate tests.

The ammonites [200] and belemnites are two fossil groups which have completely died out. The ammonites lived from the Silurian to the Cretaceous periods and there were over 600 genera and thousands of species. The larges of all was the Cretaceous *Pachydiscus septomarodensis* which measures over 7 feet across. They were heavy shelled with various ridges and ornamentations. They evolved a better swimming shape of shell than the nautiloids, with a keel which cut through the water more easily than the blunt nautiloids.

The belemnites were squid-like animals with an internal shell. They arose in the Triassic period and were all extinct by the end of the Tertiary; their size has to be judged from fossilized shells and is thought to have reached 6 to 8 feet.

The nautiloids were some 2,500 species, now reduced to three. They all had, and have, a many chambered shell external to the soft body parts. The tentacles of *Nautilus* [201] are primitive retractile filaments; the eye is simple with the advanced lens and cornea of the later cephalopods; the chromatophores are simple and there is no ink sac; all of which indicates more primitive stock.

202

PHYLUM ECHINODERMATA

The name of this familiar group of animals means spiny skinned. The skeleton lies just under the thin epidermis, and is composed of calcareous ossicles usually with projecting spines. Starfishes and brittle stars have articulating ossicles; in the sea urchins and sand dollars they become fused and form a rigid shell. The sea cucumbers have tiny ossicles scattered and embedded in the muscular body wall, and the sea lilies have some fused plates to form the stalk and calyx, and articulating plates on the arms. The larval stages are bilaterally symmetrical but on metamorphosing they become radially symmetrical with a pentamerous, or five-rayed pattern [203].

Waste materials diffuse out of the body via the gills

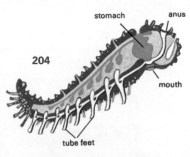

203 *Marthasterias*, Spiny Starfish *a*. larva
204 Madreporite
205 Severed arm of starfish regenerates into new starfish
206 Sea urchin : *a*. larva 7 days old, *b*. 50 days old, note growing spines, *c*. part of sea urchin showing pattern of spines
207 Echinoderm body plan *a*. crinoid *b*. holothuroid *c*. asteroid

206

and tube feet. The sexes are usually separate and the eggs and sperm are liberated into the water where fertilization may take place and give rise to free pelagic larvae. Some species retain the eggs in brood pouches. All the larvae have ciliated bands with which they swim and there are typical forms for each class.

The starfishes, or Asteroidea [207c], have a fat central disc to their body and the arms are not clearly demarcated from the body. The brittle stars, or Ophiuroidea, have a sharply defined central disc, with thin arms attached [209, 211]. The Echinoidea, the sea urchins [206], are globular or flattened discoidal forms without any separate arms. The sea lilies, the Crinoidea [207a], are rather flower-like, with the mouth held uppermost in the centre of the calyx and surrounded by branching arms. The sea cucumbers, the Holothiuroidea [207b], have partially lost the perfect pentamerous symmetry because they lie on one side; they have no arms and the body is elongated on an axis of mouth to anus, resulting in a cucumber form.

There are over 5,300 species of Echinoderms; a totally marine group, they range from abyssal to pelagic to littoral habitats. The smallest members are about $\frac{1}{2}$ inch diameter.

Asteroidea and Ophiuroidea

The starfishes vary the most from the five rayed pattern of this phylum. Many have just five arms, as the common *Asterias* and *Marthasterias* genera [203], but the arms may be increased in multiples of five. In the Sunstar, *Solaster* [212], there may be from ten to fifty arms surrounding the central plump disc.

Ceramaster [210], a cushion star, has short, blunt arms and *Culcita* has such short arms that they look just like the corners of a pentamerous figure. The starfish are carnivorous and are the scourge of oyster and mussel beds. They attach their tube feet to the bivalve and pull the shells open fractionally, they then invert their stomachs and push it into the mollusc and pour out their digestive enzymes [213]. Gradually they kill their prey and completely open the shells to finish their meal. The less

208 *Porania pulvillus*
209 *Acronida brachiata*,
Brittlestar
210 *Ceramaster placenta*,
Cushion Star
211 *Ophiura*
212 *Solaster papposus*

213 Starfish about to attack oyster

214 Gorgon's Head or Basket Star

common burrowing starfishes, like *Astropectan* and *Luidia*, cannot force open shells as the tube feet have no suckers on the end. They eat polychaetes and smaller molluscs.

Brisingia is an abyssal genus, which is rather like a brittle star in form, with many fine arms. The brittle stars have thin uniformly plated arms, which are clearly separated from the disc and do not contain any extensions of the gut. Their tube feet do not end in suckers and they move by threshing actions of the thin arms. They gain their common name by the fragile nature of the arms which break off easily when handled.

In one group of living ophiuroids, the arms can only be articulated in the horizontal plane. *Ophiothrix fragilis* is the most common brittle star of this form in the North Sea. It is found under gravel and stones, often in large numbers. *Ophiocomina* and *Ophiura* are typical members of this group, found on the surface or just buried. The other group of brittle stars are able to move their arms in all directions and form coils which can grasp algae fronds. These types are frequently found climbing and clinging to other sea life. *Gorgonocephalus* [214] is one aberrant genus; its five arms are biradiate and branching, so that they terminate in many fine coils. It is commonly called the Basket Star. *Asteronyx* and *Astroschema* are sublittoral forms which have tentacular arms. Many brittle stars, like *Amphipholis*, a common low water genus, are luminescent, emanating a yellow-green light from the base of the spines on their arms.

Echinoidea

As a class, the Echinoidea are radially symmetrical, the sea urchins being most regular, but the other orders show a tendency to revert to bilateral symmetry secondarily. No forms have arms, and the body is either globular as in *Echinus* [222], or discoidal, with the body flattened in an oral – aboral axis, as in the sand dollars. The plates of the skeleton are fused to form a complete case, the test, and this bears spines which articulate on bosses and pedicellarias, both of which may be poisonous in some species. The mouth of echinoids is equipped with an elaborate jaw system. *Paracentrotus*

215 Purple Heart Urchin
216 Piper, *Dorocidaris*
217 Green Urchin
218 Thick Spined, *Heterocentrotus*
219 *Colobocentrotus*
220 *Encope gindis*
221 Wheel Urchin
222 *Echinus esculentus* without spines to show tube feet

can actually excavate rock by persevering with the teeth and spines, gradually wearing out a hole for itself. *Asthenosoma*, from the same waters, has blue poison sacs covering the tip of each spine. Poison in these forms is used to drive away enemies and capture, by paralysis, the smaller animals.

Echinus esculentus [222] is the Common Sea Urchin, with a red-violet test covered in white bosses bearing ornamented spines. *Psammechinus* is more flattened and has a greenish shell. *Diadema antillarum*, the West Indian Urchin, is banded purple and white, whereas there are black and dark purple species of the same genus in the Mediterranean and warmer east Atlantic coast. They all have extremely long spines which are a hazard to swimmers on some beaches. *Plesiodiadema indicum* is a deep water, Indo-Pacific species. It has curious curved spines which bend down to touch the substrate, ending in spatulate tips.

The heart urchins [215], belonging to the spatangoids, have much smaller spines than the sea urchins, and are adapted for burrowing into sand. The spines are used to move the sand grains, and the tube feet are modified to help burrowing and to breathe with once buried. The Common Heart Urchin, *Echinocardium*, has a dorsal tuft of spines which hold open a breathing tube in the sand. The Keyhole Sand Dollar, *Mellita*, has six holes in the test, called lunules. The African Sand Dollar has a toothed rear margin which is used for digging. *Clypeaster* shovels sand with the teeth, and sifts out food for itself as it digs; it is a widespread genus of tropical waters.

Cidaris is a living relative of the Ordovician fossil *Bothriocidaris,* the oldest reliable echinoid fossil. The cidarids have heavy plates and spines and two rows of tube feet.

220   **221**   **222**

Crinoidea and Holothiuroidea

The Crinoidea, or sea lilies, are the most ancient echinoderm group, and they still retain some of the primitive features evolved long ago. The stalked forms, like *Pentacrinus*, *Rhizocrinus* and *Ptilocrinus*, have a stem, anchoring to the substrate with a holdfast. At the top of the stem there is a cup or calyx, surmounted by a domed roof with the mouth in the centre of this upper surface. The soft parts of the body are housed in this portion. From the edges of the calyx, there arise the arms which are feathery in that large

223 *Endoxocrinus*, Thomson's Sea Lily
224 *Halopus*, All Foot, a fixed crinoid from Barbados
225 *Antedon bifida*, Rosy Feathered Star
226 Trepang, or Béche-de-mer
227 *Labidoplax digitata*
228 Sea cucumber

numbers of pinnules branch alternately along their length. The feather stars, like *Antedon* [225] and *Notocrinus*, have no stalks in the adult form and are pelagic. There are some eighty species of sea lilies [223], and 550 species of feather stars living today. *Ptilocrinus* has the primitive structure of five arms, divided into pinnules. More often the arms subdivide until they may reach sixty as in *Metacrinus*. *Antedon rosacea* is the Common Feather Star, often dredged up from 10 fathoms round the British shores. *Antedon bifida* [225] has very delicate arms. Most feather stars such as *Neometra* are found in the warmer Indo-Pacific waters. This has thirty arms and short tentacles from the calyx base.

The Holothiuroidea, or sea cucumbers [228], include some 500 species of armless echinoderms. The tube feet are modified in the region of the mouth to form tentacles for feeding and burrowing. They are basically tube-like in form, with a soft muscular body.

Cucumaria and *Holothuria* are common species found in North America and Europe. *Holothiura* is known as the Cotton Spinner because it secretes sticky white threads from glands near the anus when disturbed. It has a leathery skin and grows usually to about 8 inches in length. *Labidoplax* [227] is a deep burrowing genus; it is extremely elongated, and leaves only the oval tentacles above the substrate with which to breathe and feed.

The sea cucumbers are eaten in various parts of the world; the Chinese call it Trepang, the French, Bêche-de-mer [226]. They may be used to prepare soups with other sea produce.

PHYLUM CHORDATA

The simplest members of this phylum have no skull, and are styled the Acrania, and those with a skull, the Craniata. The chordates without a skull or vertebral column are grouped into three subphyla – Hemichordata, Urochordata and Cephalochordata.

All chordates have a stiff axial rod, the notochord, at one stage in their development. This lies above the gut and below the nervous system and acts as a skeletal support. The nervous system of chordates is a hollow tube, the spinal cord which lies in a dorsal position, and the brain, developed from the anterior end. In the craniates, they are protected by the skull, or cranium, and the vertebral column. There is a well developed blood system and a true tail. In the craniates, there are usually two pairs of appendages, fins or limbs.

229 *Saccoglossus*
230 *Balanoglossus*
231 Lateral view of the pterobranch hemichordate, *Rhabdopleura*, in its tube
232 Diagram of body form of *Siboglinum caulleryi*

SUB-PHYLUM HEMICHORDATA

The Hemichordata include two main living types, the acorn worms and the Pterobranchiata. The acorn worms or Enteropneusta are marine burrowing creatures, rather worm-like but possessing the basic chordate features. Their body is divided into three regions, the proboscis, used for burrowing, the collar underneath which is the mouth, and the trunk which bears the numerous gill slits and gonads. *Balanoglossus* [230] is typical of the group, dwelling in mud low on the sea shore, swallowing great quantities of the substrate and extracting what food value it can.

The Pterobranchiata are sessile organisms which are rather polyp-like. They secrete themselves tubular protecting cases which are attached to the sea floor in deep waters. *Cephalodiscus* and *Rhabdopleura* [231] both have branched tentacles which are ciliated and are used for food collection. They are able to reproduce by budding; in *Rhabdopleura* [231], the buds remain attached, so giving rise to a colony; in *Cephalodiscus*, they always separate to form a new individual. A small group of animals called the Pogonophora and thought to be hemichordates are found at abyssal depths, down to 8,000 metres. One representative is *Siboglinum* [232].

232

233

a

234

SUB-PHYLUM UROCHORDATA

The sea squirts, salps and dolids are a group which have lost many chordate features in the adult form. Their relationship to the phylum is established by study of the larval forms.

The Appendicularia larva, or 'ascidian tadpole', has a trunk and a long tail, absorbed at metamorphosis. It is the tail region only which contains a notochord and a dorsal hollow nerve cord.

The Ascidiacea, or sea squirts, are sac-like animals covered with a tough case made of tunicin, a substance related to cellulose. *Ciona* [235] are common marine animals, found on seashores attached to stones or partly buried.

The Thaliacea, *Salpa* and *Doliolum* [236], and the Larvacea, *Oikopleura* [233] or *Fritillaria,* are free swimming planktonic forms. The latter group include the most specialized of all the tunicates.

235

233 *Oikopleura*
234 *Botryllus,* a compound
ascidian : *a.* larval form
235 *Ciona*
236 *Doliolum*
237 Amphioxus in
burrow : *a.* anatomy

SUB-PHYLUM CEPHALOCHORDATA

The third and last sub-phylum of the chordates without a vertebral column is one in which the basic chordate features survive through into the adult stage.

Branchiostoma or Amphioxus [237] is a very well-known animal occurring in shallow coastal waters. Tropical shores are inhabited by a related genus, *Asymmetron*; both *Branchiostoma* and *Asymmetron* are burrowing animals, living in clean sand areas. *Amphioxides* is a pelagic genus, found in tropical waters, but other than being part of the more active plankton is similar to the other genera.

The cephalochordates can swim with rapid side-to-side movements of the body wall muscles. The burrowing forms are known to make nocturnal excursions from their burrows. They are most active in the spawning season, when the separate sexes all migrate to the surface waters to shed the eggs and sperm. The embryos and larvae live as part of the plankton and then, as the larvae gradually metamorphose, they adopt the adult burrowing habit.

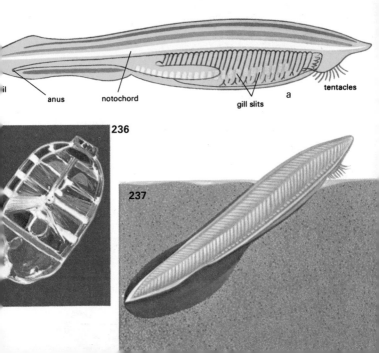

anus   notochord   gill slits   a   tentacles

236

237

SUB-PHYLUM VERTEBRATA

Cyclostoma

These are primitive fishes, each one with a circular suctorial mouth armed with pointed teeth, but lacking any supporting jaws [239]. There are two kinds of living jawless fishes, the hagfishes and the lampreys. There is fossil evidence of related groups which did not withstand the test of time, such as *Cephalaspis*.

The distinctive features of the group include an all cartilage skeleton, the simple notochord, and the numerous articulated supporting gill arches. They have a single median nostril which leads to an extremely sensitive nasal sac.

Some species are totally marine, some totally freshwater, and a few are equally at home in either habitat.

238 Lamprey: *a.* anatomy,
b. Lampreys attacking Salmon,
c. head of Lamprey
239 Fossil relatives
240 Hagfish with egg capsules

238

brain
mouth
gills
heart
stomach
notochord
a
b
c

The hagfishes [240] number about twenty species in all. *Myxine glutinosa*, the Atlantic Hagfish, holds the length record of 31 inches, though it rarely reaches half that size in European waters. All species are blind, with only eye rudiments, and have antenna-like whiskers round the mouth. When not feeding they lie in the muddy sea floor with their mouths and whiskers protruding.

The most notorious lamprey [238] is *Petromyzon marinus* which occurs round the northern Atlantic coasts, normally living in the sea, but migrating inland up to shallow freshwater streams and lakes to spawn. The eggs are laid in a rough nest, and in a few weeks hatch out into blind ammocoete larvae. The larval stage is prolonged – up to 6 years – before the metamorphosis into a full-grown, sighted, sucking-mouthed adult. The young lamprey then begins a downstream journey to the sea where it pursues a parasitic mode of life. There are about twenty-four other species of lamprey, most with life histories similar to *Petromyzon,* and a few non-parasitic species.

mouth   gill opening

241 Chimaera
242 Ray

Chondrichthyes

There are almost 600 species belonging to this class including the sharks, the skates and rays [242], and the chimaeras. Included are the largest fishes known, the Whale Shark, reaching over forty-five feet, and the Giant Basking Shark often attaining this length, both of which are filter feeders, and the most feared fish, the man-eaters and many others reputed to attack divers and swimmers.

They all possess skeletons of cartilage; paired pectoral and pelvic fins, with a variety of additional dorsal, anal and ventral fins. The skin is covered with placoid scales which arise from the basal layer of the epidermis in a similar manner to a tooth, they have roots and projecting spines. The heart is four-chambered and S-shaped, with a collecting sac, an atrium, a ventricle and a conus arteriosus, leading via the ventral aorta to the gill arteries. The jaws are made up of bars of cartilage, the upper jaw of most is attached by ligaments to the base of the cranium, but in the chimaeras [241], it is fused rigidly as in the bony fishes. The intestine is short and, internally, the surface area is greatly enlarged by a unique structure, the spiral valve. They have no swim bladder contrary to most bony fish. They all have large yolky eggs which are laid inside horny capsules, the 'mermaid's purse', or in many species, the capsule is retained inside the female while the young fish develop and they apparently are born alive. This condition is known as ovoviviparity. Internal fertilization takes place and the males are equipped with claspers, modified parts of pelvic fins, which aid the insemination of the female.

The class is further divided on the numbers and position

of gill slits which can easily be seen on the side of the head in sharks and on the dorsal surface in the skates and rays. Another bony fish feature in the Chimaeras is an opercular flap covering the four gill clefts to give a single external aperture.

The mackerel sharks include some of the world's most notorious man-eaters. They have been known to attack rowing boats and records of their stomach contents show that they are capable of consuming a whole, medium sized sealion. The Mako Shark from the Indo-Pacific and Australian waters used to be caught by Maoris, who set great store by the decorative value of the centre teeth. The hammerhead sharks [243] are perhaps the ugliest and are certainly the easiest genera to identify. Their heads are extended laterally into two flat-ended lobes, bearing the eyes and nostrils on the flat outer surface. They are recorded from many parts of the world as unprovoked attackers of swimmers and divers.

During World War II, when Norway, the chief exporter of fish liver oils, was under blockade, the Soupfin Shark liver rocketed to fame in America as a rich source of vitamin A. Fortunately, for the species, supplies were soon imported from other countries, and also, economic processes for synthesizing vitamins were developed.

243 Hammerhead shark
244 Tiger Shark
245 Sawfish

243

244

245

Osteichthyes

In contrast to the Chondrichthyes, the Osteichthyes have a skeleton which is almost entirely made of bone. The primitive members of the group still retain a cartilaginous skeleton, but have developed a cranium of bony plates. The gills are covered by a single flap, the operculum. The body is covered by an exoskeleton of bony plates, or scales, which normally overlap in a posterior direction like roof tiles. The eggs are usually fertilized externally, and therefore there are no claspers in the males. In the few instances where the eggs are fertilized internally, there is a single intromittent organ developed from the median anal fin. The air bladder, which develops as a pouch off the gut, is present in most bony fishes. In some species it acts as a lung; in the more advanced forms it becomes sealed off from the back of the pharynx and forms the swim bladder.

As an aquatic group of animals, the bony fish are extremely successful, there are over 20,000 species and the number of any one species can be enormous; it is estimated that over 3,000 million herring are caught each year so that their actual numbers must be many times more. The more ancient fishes, or Paleopterygi, include three families, the birchirs, the sturgeons and the paddlefishes, which retain certain primi-

247 Shovel Nose

tive features when compared with the fully-fledged, more modern bony fish, the Neopterygi.

The birchirs [250] are a small group found in West and East Africa. *Polyterus* is the main genus and is notable for its lobed pectoral fins, absent pelvic fins and the odd dorsal fin, which resembles half a feather.

The sturgeons [246] are found in the temperate regions of the northern hemisphere. Their skin is scaleless except for five longitudinal rows of pointed, lozenge-shaped plates. Under the snout there are four long whisker-like projections, or barbels, which sweep the floor and locate the presence of invertebrates, such as snails and worms, on which they feed.

The bowfins [249] are a very ancient group of fish whose fossil record spreads across North America and Europe. Today they are restricted to the freshwater lakes and rivers of the eastern United States. The gars [248] are restricted to North America, mainly to the east of the Rocky Mountains. They have sharp needle-like teeth, which help them catch the smaller fish on which they feed.

248 Gar Pike
249 Bowfin
250 Birchir

251 Salmon

The bulk of the **modern bony fish** are very similar in external appearance except for the different positions and numbers of fins, and whether they are soft- or spiny-rayed. Internally there is a change from a primitive condition, where all the vertebrae are similarly shaped, to a more advanced condition, where the anterior bones of the vertebral column are modified to form the 'Weberian apparatus', which is thought to respond to vibrations in the water like a kind of ear.

Among the most primitive fish are the herrings, anchovies, trout and salmon, whitefish, graylings, smetts, hatchet fishes and the mormyrids.

The first seven mentioned groups can be immediately picked out as important food fishes for man. Herring and their smaller relatives, the sardines, live in the oceans of the world,

252 Carp
253 Elephant Mormyrid
254 Pike

255 Spotted Moray
256 Cod
257 Flying Fish

and swimming in shoals, they may occur in thousands in any one patch of water. They are very nutritious fish on account of the oily flesh, which contains valuable vitamins.

The life history of the trout and salmon [251] involves the long struggle upstream from the sea to the freshwater spawning grounds. Not all members of this family toil in this manner – some are entirely freshwater, being land locked and most of the popular food varieties are now bred in hatcheries and released into lakes and streams to mature.

The hatchet fish are all small axehead-shaped fish which are most obvious by their photophores, or light producing organs. They are deep and warmer water forms, and look like luminous fish skeletons covered by transparent tissue.

The pike [254] and their relatives are less primitive than those fish just mentioned. They are all efficient carnivores, not only taking other fish, but amphibians, small birds and mammals when available. One glance at the tooth array of a Northern Pike or the American Muskellunge is sufficient to illustrate its capturing capabilities.

The eels, the cyprinid group, the catfish, the cod relatives, all the flatfish, puffers, anglers, gobies, mackerel, tunas and many others are all more advanced fish. The variety of form, colour, pattern and habitat, is indescribable. Many, such as cod, haddock, mackerel, plaice and sole, form the basis of the world fisheries and are exploited in an effort to increase the supply of protein for human needs. Many species which could be used as food are just not popular, and research as to how these species could be popularized in the protein-short areas of the world is very much needed.

256

257

255

Four families of fish seem to have developed the ability to produce an **electrical discharge** independently: the electric eels and knifefishes, the electric catfish, the mormyrids and the electric stargazers. Most of these fish live in turbulent or muddy water, where vision would be of no use in detecting the presence of prey. Muscle blocks along the side of the body have become modified into electric organs which produce an electric field around the body. If this field is interrupted by another animal, it can be located immediately and accurately, and consequently devoured.

Some of these species can produce only a very small voltage [261]. Others, for example the Electric Eel [262], produce a low voltage most of the time, but when the eel wants to immobilize any swimming object, it discharges about 600 volts, which stuns the prey and facilitates capture.

The **colour** of fishes is due to various pigments present in the epidermis, or due to the refraction of light. Pigments are contained in cells, the chromatophores, and some fish possess the ability to contract or relax their chromatophores

261

262

and alter their coloration and its pattern. The flat fish are most famous for their ability to colour and pattern match their skin to their surroundings [263]. The chromatophores seem to be under hormonal control, directed from the brain.

Apart from colour change which is restricted to only a few families, fish employ many other devices for disguise, camouflage, protection and aggression. Heavy scales prove adequate **protection** in the cartilaginous fish, the sturgeons and gars. Sharp spines cover the bodies of the puffer fish not only making them unapproachable, but also as a warning of toxicity. The scorpion fishes have dorsal fin spines which act like hypodermic needles leading from the poison gland.

Disguise is a less unfriendly form of protection. The bright colours of tropical fish act as a camouflage when set against a colourful coral setting and mock eyes on the tail often deceive attackers [258]. Temperate and arctic fish tend to blend with the sea bed or the water surface.

Camouflage is taken to the extreme in the case of the Sargassum Fish and the frog-fish and angler fish. The fins

263

258 Butterfly Fish
259 Sargassum Fish
260 Sea horses
261 Electric Cat Fish
262 Electric Eel
263 Resting flat fish shuffle sand and pebbles on to their backs. The disguise is far more complete than shown here

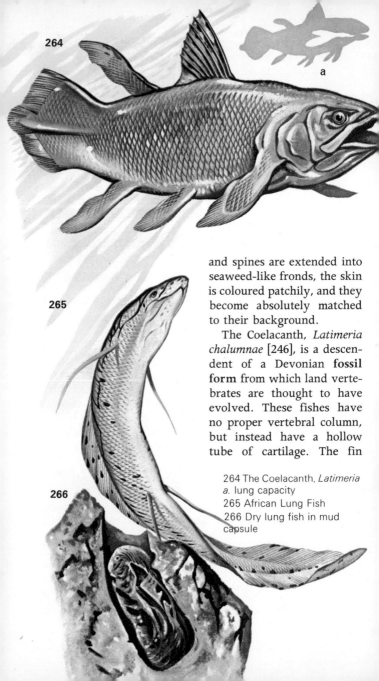

and spines are extended into seaweed-like fronds, the skin is coloured patchily, and they become absolutely matched to their background.

The Coelacanth, *Latimeria chalumnae* [246], is a descendent of a Devonian **fossil form** from which land vertebrates are thought to have evolved. These fishes have no proper vertebral column, but instead have a hollow tube of cartilage. The fin

264 The Coelacanth, *Latimeria*
a. lung capacity
265 African Lung Fish
266 Dry lung fish in mud capsule

267

spines are hollow rods of cartilage, from which the name of the group, coelacanth, or hollow spine, is derived. Another feature is the large single lung derived from the anterior end of the gut [264a]. This feature is seen in another ancient group of fish – the lung fishes. *Latimeria* has changed remarkably little from its Devonian forefathers and the Australian Lungfish, *Neoceratodus* [267], has changed little from its Triassic ancestor *Cerotadus*.

Many fishes in warmer, less well oxygenated waters, rise to the surface, **breathing** in air with the mouth and pharynx. This commonplace activity can be readily observed in aquaria. If a sheet of glass were placed over the surface, the fishes would effectively 'drown' from lack of oxygen.

The eel is known to make land journeys through damp places. The Indian Climbing Perch or Walking Fish spends most of its time on land. It walks with the spiny edges of the gill plates at about the rate of ten feet per minute.

The lung fishes have all developed a special sac as an extension of the anterior wall of the gut. It functions as a lung in that it is increased surface area for oxygen absorption. If the muddy rivers in which they live dry up, they load their lung with air and bury themselves in a cocoon of mud and hibernate until it rains [266]. The tropical mud skippers [268] can also exist out of water, as long as they can maintain a moist body surface.

268

267 Australian Lung Fish,
Neoceratodus
268 Mud skippers range from
Africa to Australia. They vary in
size from 1½ to 10 ins in a great
number of colours

269 Adult frogs mating. Comparison of male, *a*. and female, *b*. limbs

Amphibia

The most significant feature of the Amphibia, as the first terrestrial vertebrates, is the development of the pentadactyl limb. This limb is the basic pattern for all vertebrate limbs and consists of an upper part articulated to the limb girdle, a lower part hinged to the upper and terminates in a hand or foot bearing five digits in the unmodified form – hence pentadactyl. The feet are webbed and are effective for both walking on land when closed and for swimming when open.

The skin of amphibians is soft, and kept moist by mucus glands. This is essential as the skin is used as a respiratory surface. A few species, belonging to the Caecilia, retain fish-like scales embedded in the skin, but all the other species have developed an outer cornified layer to the epidermis, typical of land vertebrates. The more fully terrestrial forms have a drier, more horny skin. The epidermis contains chromatophores with three pigments, brown-black, blue-green and yellow. By combinations of these colours, the amphibians present a wide variety of colour forms and patterns. Most species can contract or expand the chomatophores and change their colour to suit their background.

The reproductive habits show least terrestrial adaptation in most species in that the eggs are shell-less and have to be laid in water or in moist situations. The larva, or tadpole, hatches out and swims freely, feeding and growing until it undergoes metamorphosis which results in the adult form [271]. Many forms lay the eggs on land but some more adapted species have egg capsules. In extreme conditions, where the adult lives entirely out of water eggs are retained inside the female and young are born as miniature adults.

The amphibians show a diversity of respiratory techniques. They all use the skin as the main respiratory surface when in water. The tadpoles and the adults of some urodeles, such as *Necturus*, the Mudpuppy, have external gills which are frilled extensions from the side of the head. Preceding metamorphosis, the gills become internal and are finally absorbed as the lungs develop towards the completion of the transition. The mouth and pharynx are respiratory surfaces, and the lungs seem to be reserved for times of greater oxygen requirement during active locomotion.

There are some 2000 living species of Amphibia, grouped into three sub-classes: the Urodela, including the newts and salamanders; the Anura or the frogs and toads; and the Apoda, a small group of tropical legless forms.

270

270 Cell division in an amphibian egg

271 Some stages in the development of the embryo of a frog: *a.* frog spawn, *b.* gilled stage, *c.* back legs, *d.* forelegs, *e.* young frog with cast skin

272 *Triturus: a.* male, *b.* female, *c.* development of larval stages

The **ceacilians** are worm-like, burrowing forms, which have undergone reduction of many features and have lost their limbs in their quest for subterranean life [273]. The head is blunt and bears the mouth, a pair of nostrils, a pair of very small lidless eyes and a pair of protrusible sensory tentacles which lie behind the nostrils. The body is long, cylindrical and marked by a series of transverse grooves which enhances their wormlike appearance. There are some fifty-five recognized species all of which inhabit the tropical regions of the world, with the exception of Madagascar.

The newts and salamanders are **tailed amphibians** that have retained more of their ancestral features than the frogs and toads [274]. The larvae and the adults are very similar to each other; they both have tails which are used as locomotor organs when swimming, and teeth in the upper and lower jaws. The more aquatic forms retain the larval feature of external gills, and always have aquatic larvae. The more terrestrial forms, like the European Salamander, have fully developed lungs and bear live young.

273 A caecilian : *a.* larva

273

The largest species is the fully aquatic giant salamander, *Megalobatrachus,* which lives in China and Japan. It reaches a length of $5\frac{1}{2}$ feet. They have flattened, warty heads and proportionately small limbs and they retain some internal larval features.

The Common British Newt, *Triturus vulgaris* [272], is more terrestrial in habit, but stays in damp vegetation, never remaining long in dry situations. Both sexes return to the water in the breeding season, the males congregating a few days before the females.

A common North American genus is *Ambystoma,* which includes eleven species ranging from Mexico to southern Alaska. They are all built to the same plan, but vary tremendously in colour and patterning. The Pacific Giant Salamander, *Dicamptodon ensatus,* is the largest terrestrial salamander, reaching a length of just over 1 foot. It inhabits the west coast of North America.

Ambystoma mexicanum exhibits a phenomenon of becoming sexually mature before all the larval features are lost if it lives in water lacking in iodine. There are many other salamanders which also exhibit this lack of complete metamorphosis, with bizarre results.

274 *Ambystoma opaculum,* Marbled Salamander *a-c.* larval stages

275 *Cryptobranchus*

In many aquatic urodeles, there is a tendency to retain certain laval features even though an individual may have attained sexual maturity. This process is called **neotony** or paedomorphosis.

The thyroid gland has been shown experimentally to control metamorphosis in the Amphibia, and iodine is necessary for the production of its hormone, thyroxin. In aquatic conditions which are deficient in iodine, the thyroid gland does not function properly and complete metamorphosis cannot take place. The gonads are not influenced by thyroxin and proceed to develop at their normal rate, resulting in apparently larval forms which can breed.

Many axolotls from North America are seen to retain the external gills, the larval finned tail and remain aquatic throughout life. This is due to lack of iodine. The Mexican Axolotl and some of the tiger salamanders from New Mexico are genetically neotonous and will not undergo metamorphosis even when treated with iodine or thyroid extract. *Cryptobranchus,* from the United States, retains a few internal larval features, particularly the internal gill arches.

Siren shows three permanent external gills, small forelimbs and has no hind limbs at all. These two genera will only partly metamorphose even when treated with iodine. *Necturus,* the Mudpuppy [277], is a North American genus; is has three large external gills, and the lungs are so reduced and non-functional that the animals need never surface for air.

Amphiuma [276], the 'Congo Eel', also from Southern

United States, is a caricature in that its body is extremely elongated with a blunt head bearing small lidless eyes. The gills are covered by an operculum, and it has very small limbs at each end of its body.

Three species of salamanders, two from the New World belonging to the same family and one European form from another family, have all adopted cave dwelling as a mode of life. Living in perpetual darkness, they have undergone various changes which are typical of such a habitat. They have lost skin pigmentation and the eyes have been reduced to functionless rudiments.

276 *Amphiuma*
277 *Necturus*

Typhlotriton [278] inhabits caves of Missouri and Arkansas and is the only one of the three which metamorphoses, to become the only blind adult salamander. *Typhlomolge* [279] is found in the caves of Texas, its colourless body bears thin extended limbs, which is another frequent feature of cave-dwelling adaptation. *Proteus* is the European Olm, found in the Balkan limestone caves. It reaches 10 inches in length and has a long white body with three pairs of pink frilled gills.

278 *Typhlotriton*
279 *Typhlomolge*

The frogs and toads form the third and last natural group of amphibians called the Anura or Salientia. These are the **tailless amphibians**. Their hind limbs are extremely powerful and act as levers enabling the animals to make long jumps or to swim with an effective kick stroke. None of the adults has gills but use the skin, buccal cavity and lungs as respiratory surfaces.

The most primitive frogs are two genera, one, in New Zealand, namely *Ascaphus*, the other in the north-western United States, *Liopelma*. The feature of their primitiveness is the presence of two tailwagging muscles.

The pipid toads are entirely aquatic; *Xenopus laevis* [288], the Smooth Clawed Frog, is often kept in laboratories. The spade-foot toads of Europe and North America are terrestrial and use a sharp projection on the inner side of the foot for digging.

280 Western Toad; the mating call is a minute-long blast
281 *Rhacophorus*, Javan Gliding Frog, 3 ins
282 The Arum Frog (South Africa) spends its life in Arum Lily flowers. Almost white specimens can be found.
283 Painted Escuerzo, *Ceratophrys ornata*

The African Burrowing Toad, *Hemisus*, is found widely throughout the arid parts of Africa. The dominant toad family is the Bufonidae, and it shows a wide range of types. *Bufo Cognatus* is a rough-skinned desert-species whereas *Bufo alcarius* is smooth-skinned and lives by streams.

The African *Astylosternus* is known as the Hairy Frog [284] because the male develops thick patches of blood-filled villosities on its thighs and sides during the mating season. *Hylorina* is a South American genus which has exceptionally long hands and feet. *Phyllomedusa* is one of the tree frogs. *Rhacophorus* [281] belongs to the Old World tree frogs and is known as the Gliding Frog.

The skin in many frogs and toads contain poison glands. The most deadly venoms are found in the skins of the Dendrobatidae and are used to tip arrows of Colombian Indians [286].

284 Hairy Frog,
Trichobatrachus robustus
285 Glass Frog
286 Three-striped Arrow Poison Frog, *Dendrobates trivittatus*
287 African Poison Frog
288 Smooth Clawed Frog,
Xenopus laevis, used for pregnancy tests by the medical profession

The toads and frogs exhibit a wide range of modifications for the **care of eggs**. *Nectophrynoides vivapara,* an East African toad, retains the eggs inside the oviduct, which functions as a uterus. There may be over 100 larvae inside the female, each with a long tail which acts as an oxygen pipeline from the mother's blood vessels.

In the aquatic South American species *Pipa pipa* [293], the male presses the eggs into the soft skin on the back of the female. Here each one develops its own little brood pouch and maintains a close contact with the blood vessels of the skin for respiratory purposes. The responsibility in *Alytes*, the Midwife Toad [290], is taken by the male, who wraps the newly laid eggs round his legs and carries them until they hatch. *Gastrotheca* species are known as marsupial frogs because they develop a pouch on the back to house the eggs. *Gastrotheca pygmaea* has two flaps of skin which fold over the eggs but do not cover them completely. *Gastrotheca marsupiata* develops a complete dorsal skin fold enclosing the eggs fully within a vascular chamber. The tree frogs *Dendrobates* and *Phyllobates* carry the tadpoles from the arboreal situation down to the streams where they can complete metamorphosis.

Males and females can be distinguished internally by the

289 Darwin's Frog ; the male has about 15 eggs in the vocal pouch
290 Midwife Toad ; the male carries eggs on its back
291 Goeldi's Frog, *Hyla goeldi*
292 Marsupial Frog with the eggs under the skin of the back

289

290

291

292

presence of either testes or ovaries, but frequently external features such as colour differences or shape of different parts of the body are sufficient. These minor external differences between male and female are called the secondary sexual characteristics. The best known examples are the nuptial pads on the inner surface of the thumb in male frogs and toads, which become enlarged and pigmented during the breeding season [269], and the dorsal crest and bright courtship colours found in the male European Newt. The functions of these changes and other less familiar ones are for attracting the females and ensuring that the eggs are fertilized in the most efficient manner.

Some male frogs and toads are heavier built than females, with much stronger forearms for gripping her sides. The villosities of the Hairy Frog [284] are not found in the female of the species. The male European Mountain Brook Newts have prehensile tails and trap passing females in its coils, the spring projections from its legbones help grip the chosen mate. The male Banana Frogs of East Africa have a formidable array of spines on the thumb and on each side of the chest.

As a group, the amphibians present a group varied in habit, behaviour and appearance, as well as being the first true land vertebrates.

293 *Pipa pipa*
294 Paradox Frog ; tadpole
a. is larger than the adult

Reptilia

Most terrestrially adapted animals began to develop from amphibian stock as far back as the Devonian period, some 350 million years ago [296]. The skin became covered by a tough layer of horny material; the eggs were laid with increased quantities of yolk and protected inside by special extra-embryonic membranes, which play an important part in terrestrial development.

On hatching from the egg, or at birth, the young reptiles are miniature replicas of the adult and do not pass through a larval stage and metamorphosis as the amphibians; they breathe by means of lungs and never have gills. Like the amphibians and fishes, they are cold blooded, but have evolved various means of preventing extremes of environmental temperature from affecting their metabolism too drastically.

Internally, they have a more advanced heart structure than amphibians. The lower jaws of reptiles consist

295 Snake skull and teeth
296 Skeleton of *Seymouria a.*
skull and pectoral girdle
297 *Uromastrix* is dark in the
morning and light in the evening

of several bones and there is only a single bone in the middle ear [295]. The teeth of reptiles show a remarkable range in shape and function. They are basically simple conical structures, without a true root; they are shed and replaced throughout life, a condition known as polyphyodonty. Snake teeth are the most modified and teeth are absent in the turtles and tortoises.

During the whole of the Mesozoic era, the reptiles increased in vast numbers and varieties but during the Cretaceous period, five out of the ten reptilian orders became extinct, particularly the dinosaurs. Of the five surviving orders, the Rhyncocephalia are represented by a single genus, *Sphenodon*; the Chelonia are well represented by the variety of tortoises and turtles; the crocodiles and alligators are a meagre relic of earlier crocodilians. The lizards, Lacertilia, and the snakes, Ophidia, are the two orders which seem to have increased their range and species after the Mesozoic era. Reptiles have few enemies outside their class; the main enemy is Man. Man kills not only for food but for skins or the venoms used as medicines.

298 *Sphenodon* has a pineal eye which reacts to sunlight

298

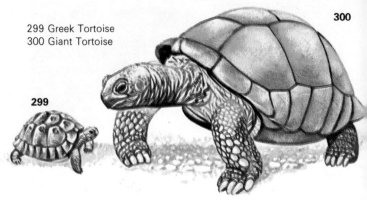

299 Greek Tortoise
300 Giant Tortoise

300

299

The Americans call all **chelonians** turtles, but in Britain, the marine and mainly aquatic freshwater types are referred to as turtles, the amphibious forms as terrapins, and the terrestrial forms, tortoises. There are over 200 species in this order and all forms are remarkably long-lived. The common *Testudo graeca* may live up to 100 years, and the giant tortoises probably live much longer.

They have all evolved massive armour plating as their instrument of survival. This shell consists of a dorsal carapace and a ventral plastron, which are usually joined laterally, leaving an anterior space for the head and forelimbs, and a posterior space for the tail and hindlimbs. It is made up of an outer layer of horny scales, bound to an inner layer of bony plates derived from the ribs and the vertebrae. As the trunk is so rigid, the neck partly compensates by being long and flexible. In most species, the head can be withdrawn completely inside the shell. The jaws are covered by a horny material to give a beak instead of teeth.

The sexes are similar, except the male has a thicker tail housing the penis, and the plastron of the male tends to be more concave to allow mounting during copulation. There is a complex courting procedure in many species, and the eggs are laid in deep burrows, dug out by the mother, usually in damp sand or earth or in rotting vegetation.

The South American Matamata, *Chelys* [302], is highly aquatic and bears fleshy tentacles on each side of the neck, which are said to lure fishes within snapping-up distance.

Dermochelys [301], the Leatherback Turtle, is the largest living turtle, reaching 7 feet in length, with a paddle span of 12 feet and weighing about 1500 pounds. It has a mainly leathery skin and just scattered bony plates embedded in it. *Chelonia*, the Green Turtle [304], is another marine turtle. It is mainly restricted to warmer waters.

Trionyx is the Soft-shelled Turtle [303], and spends much of its time in the muddy deposits by river banks. They are rather carnivorous and the back part of the beak has a special crushing surface for breaking mollusc shells. *Chrysemys* [305] is a more amphibious form, being the Common Painted Turtle from the southern United States.

Testudo is the most widely distributed genus. *Testudo graeca*, or the Greek Tortoise [299], is found commonly in southern Europe, though its numbers are sadly depleted due to its popularity on the pet market. The Giant Tortoises [300] are a different species found on the Galapagos islands, Mauritius and the Aldabra Isles.

301 Leatherback Turtle
302 South American Matamata
303 Soft-shelled Turtle

304 Green Turtle
305 Common Painted Turtle

The Lacertilia, or **lizards**, are a modern group of reptiles, and the remaining link with their ancestors is the Tuatara, *Sphenodon*, which is found on a few remote rocky islands off the coast of New Zealand. This creature is almost identical in bone structure to some Triassic fossils which were living about 170 million years ago. There are over 3000 species of lizards grouped into about 300 genera and this is therefore the largest group of living reptiles.

They are light of build compared to the other orders. The limbs are reduced in about ten families which have taken to a burrowing habit, and are consequently snake-like in appearance, but the rest have two pairs of limbs. Moveable eyelids, a

306 Chameleon
307 *Draco*, the Flying Lizard
308 Iguana
309 Gila Monster
310 Skink
311 Slow Worm

visible eardrum and small scales on the ventral surface are the features that distinguish lizards from snakes.

The monitor lizards, often called iguanas, or goannas, quite incorrectly, are properly called the Varanidae. They are an ancient group, usually reaching 4 or 5 feet in length, with elongated necks, trunks and tails, a few frills and crests.

The true iguanas [308] are New World species mainly, usually about 5 feet in length and arrayed with decorative crests along the back. *Uromastix* has a spine covered tail, and has evolved a colour change technique to simplify desert living. *Draco* [307] is a tree-dwelling agamid, which glides using the folds of skin along its flanks as a parachute.

The skinks [310] are smooth skinned lizards with short stumpy tails. These are often used as storage organs for fat which can be drawn on when food is short. They are ground dwelling and have a wide range of habitats.

The Chameleon is adapted to tree climbing by having its feet divided into inner and outer parts, two claws one way and three the other, enabling the animal to grip branches firmly. Its eyes are mounted on cones and swivel, embracing a large field of vision, and its tongue is exceedingly long and usually accurately aimed at any insect prey.

Anguis fragilis, the Slow Worm [311], is widely distributed in England, though not often seen as it burrows into soft earth or hides under loose stones. There are many relatives of the Slow Worm across Europe and Asia, for example, the glass snakes.

The only poisonous lizards are the Mexican Beaded Lizards and the Gila Monsters [309]. They come from Mexico and the deserts of North America and are typical examples of animals with warning coloration.

310

311

312 Python
313 Heat sensitive organ of the pit vipers

The Ophidia, or **snakes**, are perhaps among the most feared animals in the world, though many are completely harmless. They are distinguished from the lizards by the drastic reduction of limbs and limb girdles, the absence of movable eyelids, the loss of an external eardrum, and a single row of wide overlapping scales on the ventral surface.

The boas and the pythons [312] are large constricting snakes whose prey is usually killed by suffocation. The boas are mainly New World and the Pythons Old World.

314 Cobra
315 Anaconda
316 Oviparous birth
317 Viviparous birth
318 Sea snake

316

317

The Anaconda [315] is a water loving species and is rivalled
for size by the Reticulated Python of Malaya, both reaching
about 30 feet in length. The boas are viviparous in that they
give birth to live young [317], they may have a brood of
thirty baby snakes at one time. The pythons lay eggs [316].
The colubrid snakes are a large group containing over half
the total species of snakes. Many are harmless, but the highly
poisonous and deadly accurate biters are front fanged with
hollow channels to carry the venom into the pierced flesh.
These include cobras [314], mambas, kraits and coral snakes.

The sea snakes [318] are found in the surface waters of the
tropical seas of Burma and South China, though *Pelamis
platurus* reaches the west coast of South America. They all
feed mainly on fish and are poisonous and front fanged.

The viperid snakes are the most highly specialized
venomous snakes. The fangs are hollow and can be articulated
into an erect position for a strike. The vipers are an Old
World species, including the Puff Adder and the Gaboon Viper.
The Crotalinae or pit vipers are the New World species,
which are unique due to the presence of a deep pit, a heat
sensitive organ, in front of the eyes [313]. *Crotalus*, the
Rattlesnake, has such an organ, as has the largest viper, the
12 feet long Bushmaster from the tropics of South America.

318

319 Mississippi Alligator

Crocodilians are familiar creatures. They have elongated snouts, comparatively heavy trunks and powerful tails, which are the main organs of locomotion, being lashed from side to side during swimming [320]. The forelimbs are slightly smaller than the hindlimbs, and both are used for walking on land, though this is a somewhat cumbersome process. The teeth are conical and sharp, set in rows along the edge of the jaws and used for grasping and killing food. The eyes are set high up on the head as an adaptation to aquatic habit. The nostrils are placed on the top of the snout and can be closed for submersion. Similarly the eardrum is covered by a scale-like flap which protects it from damage. The skin is hard and scaly, reinforced by bony plates in the dermis on the back and in the tail.

Many species construct rough nests of broken branches and leaves. These can be 6 feet across and form a prominent mound on top of which the eggs are laid. The female may guard the nest until they hatch, which takes between two and three months. The Nile Crocodile buries her eggs in

320 Crocodile

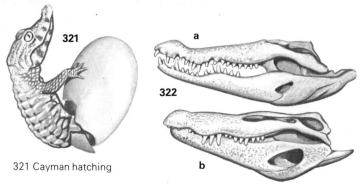

321 Cayman hatching

deep holes in the sand. As they are cold blooded, the eggs do not have to be incubated. When the young are ready to hatch, their croaking cries are heard by the mother who uncovers them. They hatch by means of an egg tooth, also found in young birds, which is used to break through the shell. This disappears soon after hatching.

The true alligators live in the Mississippi Basin [319], but one species, *Alligator sinensis*, lives in the rivers of southern China. The Caymans [321] are closely allied to the alligators and are found in South America.

The Gharial, *Gavialis gangeticus* [324], inhabits northern India. The mature males have a swollen tip to the end of their otherwise narrow snout and is the only example of an external difference between sexes in the crocodilians.

Crocodilus niloticus, the Nile Crocodile, has a long wide snout. It is restricted to freshwater, whereas the Esturine Crocodile, *Crocodilus porosus* [323], is at home in the brackish water and consequently has a wider distribution, being able to spread along shores to new regions.

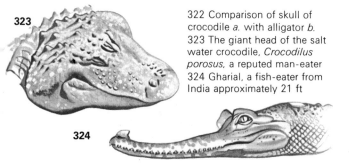

322 Comparison of skull of crocodile *a*. with alligator *b*.
323 The giant head of the salt water crocodile, *Crocodilus porosus*, a reputed man-eater
324 Gharial, a fish-eater from India approximately 21 ft

Aves

The earliest birds evolved from flying reptiles which had leathery wings capable only of gliding flight. The earliest fossil record is that of Archeopteryx which had feathers similar to present-day birds, and a long tail, still very reptilian in size, but also edged with feathers.

Feathers are grown from follicles in the skin, in the same way that mammalian hairs are, but they are different in their final structure. They do not grow continuously as hair does, but if the feather is lost by moulting or in a fight, then the papilla at the base of the follicle is stimulated to grow a new feather [332]. They are light, flexible, and provide the streamlined contours for flight. They are also waterproof and insulated due to the dead air spaces which they hold next to the skin. Feathers also play important roles in species and sex recognition and are used in courtship displays.

In flying birds, the feathers are barbed and stiff, forming an aerofoil, but the feathers of the flightless birds are soft and downy [328]. The lungs of birds are extended into extra air sacs which lie in air pockets in the neck and trunk bones. All the bones are modified for flight; the forelimb forms the wing with the reduction of the digits; the sternum and clavicles support the strong pectoral flight muscles; the keel being an extended surface area for muscle attachment. The ribs are rigidly attached to the vertebral column and sternum, unlike the mammals [327]. The wings in flight act as the wing and the propeller of an aeroplane at the same time,

325

326

325 Skeleton of a bird
326 Enlargement to show the light structure of bird bone
327 Anatomy of wing
328 Wing of Kiwi
329 Semi-plume feather
330 Quill feather
331 Magnified view of barbules
332 Development of a feather

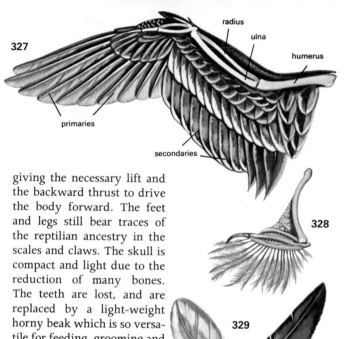

327

radius
ulna
humerus
primaries
secondaries

328

329

330

331

giving the necessary lift and the backward thrust to drive the body forward. The feet and legs still bear traces of the reptilian ancestry in the scales and claws. The skull is compact and light due to the reduction of many bones. The teeth are lost, and are replaced by a light-weight horny beak which is so versatile for feeding, grooming and defence that it leaves the forelimbs free to function as wings. Young birds have a single tooth in front of the upper beak to help break the shell, but this disappears after one or two days.

The constant body temperature, ranging between 105° and 111°F in different species, is much higher than that of mammals. As birds do not sweat, they control their body temperature by panting and after flight, a great amount of internal heat has to be lost along with water vapour from the lung and air sac surfaces.

papilla or outgrowths of skin
pulp
developing sheath
barbules

332

333 Emperor Penguin
334 Cassowary

335 Rock Hopper Penguin
336 Mantell's Kiwi

The **flightless birds** show more primitive skeletal features than the flying birds. The sternum keel is absent in the ratite, or primitive flightless birds, and the wing bones are reduced as in the ostriches. In the more modern groups, which have become flightless as a secondary feature, the traces of the keel remain. The penguins [333, 335] use their wings to row through the water in a flight-like motion, and the keel is retained for the pectoral muscle attachment. In running forms, the hind limb is powerful enabling a high speed to be attained. The feathers act simply as an insulating layer.

The Ostrich is the largest living bird, with enormous legs and thigh muscles, standing some 8 feet high and weighing over 300 pounds. They are now found only in Africa and south west Asia. The South American Rhea and the Australian Emu [337] and Cassowary [334] are close relatives of the Ostrich and are a good example of discontinuous distribution of a once widespread order.

The kiwis of New Zealand [336] are smaller animals, quiet,

337 Rhea

337

338

338 Extinct Moa

terrestrial and night feeders. They have tiny eyes hidden under fringing feathers and a long pointed beak. Penguins are specialized for aquatic life with webbed feet as well as flipperlike wings. They are found in Antarctica and southern hemisphere islands as far north as the southern tip of South Africa.

The Moas of New Zealand [338] were even larger than Ostriches and are estimated to have been 10 feet tall. Their extinction was due to hunting by the Maoris. The Dodo [339] suffered a similar fate on the island of Mauritius as recently as 300 years ago.

339 Extinct Dodo

339

The diverse adaptations of the **beaks and feet** of birds is remarkable evidence for the influence of the environment in shaping the basic structure. The beak and tongue are most modified for feeding, and the feet, although used by some species to capture food and hunt, are mainly adapted to different types of swimming, walking or perching habits.

The birds of prey, including the hawks, kestrels, eagles, buzzards and vultures are day hunters, and have powerful claws or talons for grasping prey. Their beaks are sharply curved for ripping flesh. Swimming birds have paddle-like feet, either with the toes completely joined by skin to form a webbed foot, such as in the Pelican [341], or each toe individually lobed as in the grebes.

When the ground is marshy and unstable, the feet develop extended length to distribute the weight and prevent sinking. The Lilly-trotter has extremely long toes to enable it to walk

340 Beak structures: *a.* Hornbill, *b.* Flamingo, *c.* Curlew, *d.* Cockatoo, *e.* Jaçana, *f.* Woodpecker, *g.* Fish Eagle

over aquatic vegetation. Elongated claws are also a good aid for getting a better grip on soft ground. Feathered toes act as snow shoes in the Ptarmigan's case [353].

Over half the known birds are perching in habit, with four toes arranged to give a grip as they perch. The woodpeckers have two toes forwards and two directed backwards, with sharp claws for climbing up the tree surfaces [342].

Short thin bills are found in insectivorous birds; these may be slightly curved as in the case of the tree creepers. Woodpeckers are less delicate feeders and have strong bills for hammering off the bark to reach the insect life beneath [340].

Sifting is a popular mode of feeding in aquatic birds. The Flamingo filters off planktonic crustacea in its ungainly shaped beak. Fish catchers, herons, kingfishers and many seashore dwellers, have bills with fine serrations to help grasp their slippery meal.

341 Pelican showing inside of its enormous pouch
342 Feet structures: a. Flamingo, b. Curlew, c. Cockatoo, d. Hornbill, e. Jaçana, f. Woodpecker, g. Fish Eagle

As a class, birds have remarkable patterns of **behaviour**. *Geospiza* and *Camarhynchus*, two of Darwin's finch genera from the Galapagos Islands, show how the food available in the various habitats on the islands has favoured the evolution of specially adapted bills; thick-based, tough bills for seed cracking; fine, pointed for insect picking; long and probing for eating cactus flowers and fruit.

Many birds occupy clearly defined territories which they defend against their own species by fighting, singing and display in terms of definite song patterns, and threatening posture. The main function of defined territories is that territory initiates and, by its very

343 The Woodpecker Finch uses a twig or thorn to prod out insects or grubs from crevices in the bark

344 Herons stand guard over their tree-top territories

345

346

345 The Arctic Tern makes an annual migration of 22,000 miles. Map shows northern nesting areas and main routes to the south.
346 The colourful Waxwing comes south in winter from Scandinavia and the far north.

concept, causes the evolution of behaviour patterns which maintain the mating pair.

Social behaviour is exhibited in hens, pigeons, Jackdaws and other birds which live in communities, some more close knit and involving more complex behaviour patterns than others. A 'pecking order' or rank is quickly established, and constantly changing, in such communities [347].

Migration is extraordinarily complex, and involves the seasonal flight of hundreds of birds from one area of the world to another, usually to the temperate breeding grounds during summer. The Arctic Tern breeds on the American coast, and flies to the Antarctic via Europe and the eastern Atlantic shores, making what must be the longest migratory flight of all birds [345].

347 Pecking order in Jackdaws: *a.* lesser rank *b.* top rank. This should not be confused with aggressive or courtship display.

347

Courtship in birds is the means whereby pair bonds are built up between males and females in order to rear the young as a parental team.

In the cock pheasants and peacocks, the beautifully patterned plumage and tail fans are obvious courtship displays. The birds of paradise [351, 352] are famous for their fantastic feather displays, long tail plumes, neck ruffs and crests. The Crested Grebes [348] court each other with mutual head shaking, presentation of food, and a variety of bending and rearing postures. The bitterns are equipped with a serrated claw they use for preening [350a].

The **nest** of birds is the cradle for the eggs and acts as a temporary home for the fledgelings, as they are helpless when born, until they become independent. Nests

348 Courtship of Crested Grebes
349 Bittern freezing
350 Bittern removing eel slime with toothed claw, a.
351 Great Bird of Paradise
352 Lyre Bird. The female a. also displays
353 Two Ptarmigan
354 The gull regurgitates food for its chick

a

vary from a hollow in the sand on the seashore, perhaps lined with a few old pieces of straw and feathers, to the complex interwoven and knotted fibres of the weaver birds and the carefully stitched nest covers of the tailor birds.

The eggs of birds vary in size, shape, colour, numbers laid in each clutch and the time taken for hatching. The yolk and albumen quantities reflect the stage of development at hatching, and hence give some clue to how much parental care is necessary. Most birds will lay a second clutch if the first fails to hatch or the brood perish at the hands of a predator. The fledglings are warmed, fed and watered; their faeces are cleaned out of the nest, they are later taught to seek food and to fly, accompanying the parents on excursions from the nest.

Instinctive behaviour plays a large part in most of these processes [349]. The young gulls peck instinctively at the red patch on the mother's beak; this pecking action in turn causes regurgiatation of the food that the mother has just collected and so the feeding of the young is ensured [354].

353

354

Mammalia

Birds and mammals are the only two classes of animals which have developed all the physiological and structural means of maintaining a constant warm body temperature. This faculty has meant that a wider variety of habitats have been tolerated and mastered by these groups which were not open to reptiles and amphibia.

The epidermis of mammals is covered with hairs, which grow all the time [356]. This hair, or fur, holds an insulating layer of air next to the skin, helping to prevent heat loss from the body. When a mammal wants to loose heat it can pant or sweat or increase the blood supply to the skin surface. The skin is waterproof and is an important sense organ. Apart from the Monotremata, all mammals retain the eggs inside the womb. They are fertilized internally and grow sheltered and nourished by the mother until they are ready to be born. At birth, most mammals are completely helpless, they are fed on milk from the mammary glands of the mother, and undergo a period of parental care. The other factor which has led to the success of the mammals is their curiosity, tempered with caution and ingenuity which enables them to make the best of available conditions and if necessary to seek new habitats.

Monotremes are different from all other mammals in that they lay leathery shelled eggs, instead of retaining the embryo in the womb. They have mammary glands, but without nipples from which the young can suck easily, so the mother has to let a pool of milk gather on her upturned belly and the hatched young lap it up like kittens.

355 Echidna

355

hair

epidermis

sweat gland

dermis

356 Section through mammalian skin

The monotremes as a group are restricted in distribution to Australia, Tasmania and New Guinea, and are divided into two main living families, the echidnas, or spiny anteaters, and the Duck-billed Platypuses.

The Platypuses [357] have webbed feet, sleek fur and feed underwater with their shovel-like bills on a variety of molluscs and crustaceans. They build burrows in river banks, above the water level, but with a submerged entrance. The eggs are incubated in a nest chamber inside the burrow and the young take to the water when they are about 8 inches long. The echidnas number about four or five species [355]. The Australian Spiny Ant-eater looks rather like a hedgehog, with stout spines on its back. It has a fine tapered snout, through which it protrudes its tongue when licking up ants. The Long-beaked Echidna from New Guinea is rather larger, less spinose, and has a much longer snout. Its legs hold it clear from the ground and it has enormous claws on its hind feet, which curve outwards from the body for burrowing.

357 Platypus

357

358

The pouched mammals, or **marsupials**, are very similar to the true placental mammals but the embryo is not attached to the wall of the uterus, except in a few species where the yolk sac forms a temporary anchorage. The embryoes are born in a very underdeveloped state; in the opossum only 8 days after conception and only 39 days after conception in the kangaroo. They crawl from the vulva, through a track in the mother's fur, into the pouch and attach themselves to the teats of the mammary glands. The lips grow over the teats and milk is squeezed out by special muscles of the glands down into the throat of the embryo so that it can breathe and feed without choking. The young may remain in the pouch for several months and, even when they are capable of walking and feeding on their own in the

359

358 *Didelphys virginiana*, a mother with young
359 Opossum babies, 4½ weeks old
360 Wombat
361 Marsupial Mole
362 Koala
363 Pouched Wolf or Tasmanian Wolf
364 Rat kangaroo of which there are 24 different species

360

361

362

outside world, they frequently return to the pouch as a home. There are some 150 species of marsupials in Australasia and two families of marsupials in the Americas. Commonly called oppossums [359]. *Didelphys*, the American Oppossum [358], is an inhabitant of dense woodland. It grows up to 3 feet long, including the furless prehensile tail and feeds on vegetables, fruit, small mammals and eggs.

The Australian marsupials show a wide variety of feeding, locomotory and structural types. The carnivorous marsupials include the native cats and the Tasmanian Devil and even flesh-eating mice. The Tasmanian Wolf [363], *Thylacinus cynocephalus*, is the largest of the carnivores. It is dog-like, with tiger stripes over its haunches and back. They are very rare on the island today, because they were a menace to sheep and poultry. The Marsupial Mole, *Notoryctis* [361], is a blind, burrowing mole, which feeds on grubs in the soil.

The Koala [362] is a bear-like marsupial which is entirely arboreal. It lives in the eucalyptus trees and feeds on its leaves from whence it gets enough moisture. The wombats [360] are restricted to southeast Australian forests. They are burrowing, nocturnal animals, with rodent-like front teeth. The Rock Wallaby is the most agile marsupial and the Red Kangaroo is the largest, reaching well over 6 feet in height.

363

364

The rest of the class Mammalia, the **placental mammals**, are more familiar animals; all retain their young in the uterus, developing a placenta from the embryonic tissue to the wall of the uterus, drawing food and oxygen from the mother, and having a water sac around the embryo for protection. They all undergo a long period of parental care, which is usually relative to their life expectancy.

True **flight** is found only in one order of mammals, the bats. Gliding flight has arisen independently as a method of locomotion in two other orders, the Dermoptera, or Flying Lemurs, and twice in the Rodentia with the Flying Squirrels and the Scaley-tailed Flying Squirrels.

The Chiroptera, or bats, have forelimbs developed into a new type of wing for the animal kingdom. The unique modification is that of the hand, where the first digit is reduced almost to a claw and the other four digits are greatly elongated with a

365 Fruit Bat or Flying Fox
366 Long-eared Bat: *a-b*. method of folding the ears. First one is raised and folded under the wing, *a*. then the other, *b*.

368

naked membrane of skin stretched between them and attaching them to the side of the thorax.

The fruit bats are long snouted and entirely Old World in distribution, commonly found in Malaysia, Africa, North Australia and in the tropical forests. The Malayan Fruit Bat, or Flying Fox [365], is the largest bat, with a wing span of 5 feet. The insectivorous bats have shorter snouts, better teeth and are more widely distributed than the fruit bats in temperate and tropical regions of the world. Some thirteen species of this, the second largest order of mammals, are found in Britain. The Long-eared Bat [366] has disproportionately large ears and is commonly found hunting at night.

The scaley-tailed and true flying squirrels [367, 368] are both rodents which have developed a fold of skin between the arms and legs on the lateral trunk wall, sometimes continuing to the tail. When the animals jump from a height, the skin balloons upwards like a parachute and slows down the falling speed. The American Flying Squirrel is about 1 foot in length and ranges over North American woodlands.

The Flying Lemurs are grouped together as the order Dermoptera, and have a much better flight fold, or patagium, than the previous two groups. They are found in the tropical forests of the Philippines and Malaysian islands, and are able to glide for over 200 feet [369].

369

367 *Glaucomys volans*,
Eastern Flying Squirrel
368 *Glaucomys sabrinus*,
Northern Flying Squirrel
369 Cobego

There are two orders of **aquatic mammals** that spend their entire life in the water; these are the sea cows, or Sirenia, and the whales and dolphins, or Cetacea.

Three families of the order Carnivora have become variously adapted for swimming and remain on the land for a good deal of the time; these are the seals, sealions and walruses. In a fourth carnivoran family, the Mustelidae, the otter has taken to the water to hunt for food. The rodents have several aquatic genera; the Beaver [373], Musk Rat, and Coypu all swim occasionally when foraging. These latter types, which are not entirely aquatic, have evolved structures which enable them to swim, yet still retaining the ability to run quickly on land, and maintain their warmth due to fur.

The fin-footed carnivores are much better adapted to sea life. They have developed the use of blubber, or thick fat layers in the dermis, to help insulate their bodies. The walruses [370] have lost most of their hair but the seals and sealions [393] retain a short hair coat. They all have nostrils and ear-holes which can be closed up during diving to prevent drowning and damage to the ear drum. The sealions

371

372

have a small external ear, or pinna, which distinguishes them from the seals. The walruses have enormous tusks, the canine teeth, which may be used as weapons and to dig up shellfish. The Sirenia are streamlined, hairless mammals, apart from their whiskers, whose forelimbs are like flippers and whose hindlimbs form a single large tail-like paddle. They are found along the tropical seashores and estuaries of the world and include the manatees [372] from the Amazon mouth, the West Indies and Florida, and one from West Africa, and dugongs from Arabia, Indian and Australian coasts. They all have the teats between the forelimbs, so that the mother can float in the water and suckle her young held between the forelimbs.

The cetaceans are even more streamlined, and are powerful swimmers. Blubber is their only insulator, as they have lost their hair to aid swimming. The forelimbs form stiff flippers used for steering and balancing and the hind limbs are reduced to vestiges. The ears are small openings, blocked with wax to prevent damage. The nostrils are moved up to the top of the head, forming the blow hole, through which air is breathed. The male Narwhal [374] has a spiral tusk in its upper jaw to which no definite function has been attributed. The dolphins are famous for producing sounds in the chamber between their mouth and blowhole, with which they communicate and this helps in the survival of the species.

370 Walrus
371 Otter
372 Manatee
373 Beaver

374 Narwhal with the horn growing from the upper left side of the jaw
375 Dolphin

376 Camel
377 Arctic Fox

The presence of hair is one of the primary features of mammals. In its normal condition the hairs grow densely and form a thick insulating fur layer. There are many **variations of the natural covering**. In warmer climates, the hairs do not have to protect the animal from cold quite so much so that short fur is more in evidence. In arctic conditions the hair is long. The Polar Bear has 3 inch long hair, the densest hair of all mammals. Hair even grows between the toes, which gives a grip on snow as well as preventing heat loss.

The elephant, with its enormous mass, has problems in getting rid of heat and so is relatively hairless, especially when compared with its woolley mammoth ancestors of the Ice Ages. Animals which live in climates with marked seasonal variation in average temperature alter their coats accordingly. The camel [376] has a short sparse summer coat and a thick shaggy winter coat. In the north temperate regions, where there is snow on the ground in winter, many mammals grow a winter coat of white fur to aid concealment as they would

378 Porcupine
379 Armadillo
380 Pangolin

be very conspicuous if they retained their normal brown or black coats. The temperature and day length changes are thought to induce this growth of white winter fur and it is seen in the Stoat, which becomes the regal ermine, and in the Arctic Fox [377], whose summer coat of grey and brown changes to pure white.

The primates have tended to lose hair on the feet, palms and face, especially Man, who has reduced hair length and distribution. The coloured patches of naked skin in the drills, mandrills and baboons are used for species recognition and communication. The faces of the anthropoid apes and Man are less hairy and this is thought to be correlated with the increasing importance of facial expression.

Horns and antlers, nails and claws, are specialized parts of the natural covering of mammals. Horns are made of keratin in rhinoceroses, as are the nails and hairs of every mammal. The antlers of deer are bony growths covered with skin, and are living until the blood supply is cut off.

The armadilloes [379] have a covering of scales, partly formed from the keratin layer of the skin and partly by bony plates developed in the dermis. This amounts to a skeleton under the epidermis and they are the only mammals with this odd feature. The pangolins [380] have a hairy ventral surface, and the dorsal surface is completely covered by large overlapping horny plates. Porcupines [378] and hedgehogs have quills which are really just stiff hairs.

381 Giant Ant Eater
382 Elephant Shrew
383 Capybara
384 Aardvark
385 Giant Panda

Most mammals are equipped with teeth adapted to their **mode of feeding** and the nature of the food, that is, hardness, digestibility and how much it has to be chewed.

Herbivores, feeding on grass, need flat edged incisors that snip the blades. Many lose the incisors in the upper jaw and develop a hard pad to press the lower incisors against. They need a large grinding area to break up the cellulose cell walls of plants. Leaf eaters need tongues and lips which can gather material into the mouth. Giraffes [389] use their long tongues to curl round leaves. The Black Rhinoceros [387] has an upper lip drawn out into a point and there are no incisors or canines. It browses on leaves and shoots and grinds them with its flat-topped premolars and molars. The White Rhinoceros [386] has a broad mouth front for grazing on herbage.

The Carnivora as an order are equipped with incisors which are mainly used for grooming or occasionally for biting; canines which are sharply pointed and used for killing and stabbing; premolars and molars which are ridged with sharp cusps. These act as scissors when the jaws close. The cats and dogs have typical carnivore dentition, swallowing their meat without chewing. The Giant Panda [385] is not carnivorous although one of the Carnivora, and uses its canines for splitting bamboo shoots.

The Insectivora are more primitive mammals and have a tooth row of unspecialized conical teeth which are suited to crushing insect bodies. The Elephant Shrew [382] has a particularly long snout and tooth row. The anteaters [381] and tamanduas have no teeth but a long drawn out snout in the form of a tube and a long sticky tongue for gathering up insects. The Aardvark [384] feeds in a similar way.

The rodents are readily recognizable by the curved chisel like front incisors, lack of canines and ridged back teeth. The Capybara [383], living in the marshy areas of South America, frequently uses its incisors for cutting down sugar cane and is the largest rodent in the world.

386 White Rhinoceros
387 Black Rhinoceros

387

386

388 Gibbon
389 Giraffe
390 Chimpanzee

Locomotion in mammals is affected by the external factors which govern all animals; gravity is the force which exerts pressure and work has to be done against the surrounding medium in order to bring about movement; mammals in water have to push against it in order to propel themselves forward. Terrestrial mammals push against the substratum and flying mammals have the most difficult problem of moving in the relatively rare medium of air. Mammals in water do not have to support their own body weight, which is why they can reach the vast size

388

389

390

391 Three-toed Sloth suspends itself with its four limbs from long hooked claws

391

they do. The Blue Whale is the extreme, weighing over 150 tons, compared with the largest living terrestrial mammal, the African Elephant, weighing up to 6 tons. Swimming mammals are streamlined. They develop flippers or webbed feet to push against the water [392]. The sealions [393] are least removed from the terrestrial animals and are still quite agile when using their flipper legs.

On land, the first problem which every mammal has to face is supporting itself on two or four legs. Most mammals use four legs, that is they are quadrupedal, and they can walk, trot, pace and gallop, using the legs in different patterns [389]. Life in the trees has two possibilities; either to run along the tops of branches, or to use the hands and arms as grasping and swinging organs. The gibbons [388] are probably the fastest movers, not even gripping the branches properly. In comparison sloths are far from active [391].

The apes are capable of walking on the hind legs only for short distances [390]. The bipedal habit has been exploited only by the hoppers like the kangaroo, which relies on its tail for balancing, and by Man.

393

392 Common Seal
393 Sealion

392

It is thought that the ancestors of the tree shrews of South East Asia first led the way to arboreal dwelling and from these small beginnings the **primates and Man** evolved. The brain in primates is large and well developed, and the eyes are placed frontally to give binocular vision necessary for arboreal life. The limbs are long and the five digits of the hands and feet are extremely manipulable.

The lemurs [395] are found only in Madagascar and have fox-like faces. The tarsiers [394] are nocturnal, with enormous eyes and pads on the ends of their digits to help climbing and jumping. The marmosets [397] are entirely New World primates, forest dwellers, and tend to be herbivorous in diet. The spider monkeys are also New World inhabitants, distinguished from the Old World monkeys by their prehensile tails. There are many species of Old World monkeys including the Drills, Mandrills and baboons. The gibbons [388] are the light-weight apes, brachiating in agile troupes through the tropical forests of Indonesia and South East Asia.

The Chimpanzees [390], Orang-utans [399] and Gorillas [400] are the heavier apes. They are intelligent animals with some patterns of behaviour similar to our own and it is a short step from ape ancestor to a Man ancestor when the rest of the animal kingdom, in all its variety, puts this step in its true perspective.

399

400

GLOSSARY

A few of the less common terms used and not fully described within the text:

Capillary: minute tube conveying blood
Carapace: shield of exoskeleton of anthropods or chelonians
Carnivorous: flesh-eating
Chaetae: bristle, usually composed of chitin
Chromatophore: cell with pigment in its cytoplasm
Cilium (pl. **cilia**): fine thread projecting with others from a cell surface
Coelom: main body cavity in which gut is suspended
Ectoparasite: parasite on outside of another organism
Ectoplasm: outer layer of cytoplasm differing from inner layer **endoplasm**
Endoparasite: parasite living within another organism
Enteron: body cavity
Errant: moving
Exoskeleton: skeleton covering the outside of the body or in the skin
Flagellum (pl. **flagella**): long thread projecting from a cell
Gonads: gamete-producing organ
Haemoglobin: oxygen-carrying pigment of red blood cells
Metamorphosis: transformation period from larva to adult
Nematocyst: a thread cell with specific function
Oviparous: laying eggs in which embryos are little developed
Pathogenic: causing disease hence **non-pathogenic**
Pelagic: inhabiting a mass of water as distinct from sea or lake bottom
Polymorphous: of many different forms
Pseudopodium (pl. **pseudopodia**): temporary protrusion of cell wall associated with flowing movement of cytoplasm
Septum (pl. **septa**): partition
Sessile: directly attached
Sporozoites, spores: multiple fission produced by asexual reproduction
Viviparous: having embryos which develop inside the maternal organism
Zygote: the fertilized ovum before division

INDEX

SOME OTHER TITLES IN THIS SERIES

Natural History

Australian Animals
Bird Behaviour
Birds of Prey
Fishes of the World
Fossil Man
A Guide to the Seashore
Life in the Sea

Mammals of the World
Natural History Collecting
The Plant Kingdom
Prehistoric Animals
Snakes of the World
Wild Cats

Gardening

Chrysanthemums
Garden Flowers

Garden Shrubs
Roses

Popular Science

Atomic Energy
Computers at Work
Electronics

Mathematics
Microscopes & Microscopic Life
The Weather

Arts

Architecture
Jewellery

Porcelain
Victoriana

General Information

Flags
Military Uniforms
Rockets & Missiles
Sailing

Sailing Ships & Sailing Craft
Sea Fishing
Trains

Domestic Animals and Pets

Budgerigars
Cats
Dogs

Horses & Ponies
Pets for Children

Domestic Science

Flower Arranging

History & Mythology

Discovery of
    Africa
    North America
    The American West
    Japan

Myths & Legends of
    Ancient Egypt
    Ancient Greece
    The South Seas